Participatory Ergonomics

Kageyu Noro and Andrew Imada

Participatory Ergonomics

Edited by

K. Noro[1] and A. S. Imada[2]

[1]*Waseda University, Tokorozawa, 359 Japan*
[2]*University of Southern California, Los Angeles, USA*

Taylor & Francis
London • New York • Philadelphia
1991

UK   Taylor & Francis Ltd, 4 John St., London WC1N 2ET

USA   Taylor & Francis Inc., 1900 Frost Road, Suite 101, Bristol, PA 19007

British Library Cataloguing in Publication Data
Participatory ergonomics.
   1. Organizations. Management. Applications of ergonomics
   I. Noro, K.   II. Imada, Andrew S.
   658.4

   ISBN 0-85066-382-2

Library of Congress Cataloging-in-Publication Data
Participatory ergonomics/edited by K. Noro and A. Imada.
   p.   cm.
   ISBN 0-85066-382-2
   1. Human engineering—Decision
making.   2. Management—Employee participation.
I. Noro, Kageyu.   II. Imada, A. (Andrew)
TA166.P37   1991
620.8′ 2—dc20   90–10834
                                                       CIP

Phototypesetting by
Chapterhouse, The Cloisters, Formby, England.

Printed in Great Britain by Burgess Science Press, Basingstoke on paper which has a specified pH value on final paper manufacture of not less than 7.5 and therefore 'acid free'.

Contents

Preface

This book describes different methods for introducing and implementing ergonomics. The first two chapters overview the rationale, value, tools and guidelines for using this participatory approach. This is followed by eight specific applications of participatory ergonomics across a wide variety of settings, problems, and cultures. Participatory ergonomics is intended for nonexperts. Therefore, the approaches and methods differ greatly from ergonomics intended exclusively for experts. Participatory ergonomics is not intended for factory workers alone. In recent years, participatory techniques have been expanded into unexpected fields. This expansion has been necessary to meet the changing needs of technology, organizations, and people. The specific applications of this approach demonstrate how participation can be used to enhance the positive effects that ergonomics can have.

The concept of participatory ergonomics originated from discussions between Dr Noro and Dr Kazutaka Kogi in Singapore in 1983. The term 'participatory ergonomics' was proposed by Dr Kogi in a subsequent discussion (Noro and Kogi, 1985). The concept was further solidified in a workshop held by Dr Noro (Noro, 1984) in Toronto. Our joint work began with informal discussions in Sacramento, California in 1984. This was followed by a workshop we did together at the first ODAM meeting in Hawaii later that year. Dr Noro proposed the term 'participatory approach' or 'participatory ergonomics' for the Hawaii workshop (Noro and Imada, 1984).

At the IEA Congress in Bournemouth in 1985, R. Sell and L. K. Martensson joined Kogi, Imada and Noro in a panel discussion (Sell *et al.*, 1985). The participants were apparently impressed by Dr Kogi's report on the actual situation of participatory ergonomics in Asia. At a workshop at the second ODAM symposium in Vancouver in August 1986, Professor Nagamachi joined Noro and Imada to add to the variety of the workshop (Imada *et al.*, 1986). At a Japanese ODAM symposium at Ryukoku University in August 1987, Dr Liker of the University of Michigan reported on participatory ergonomics in the USA. Imada chaired a subsequent panel discussion on participatory ergonomics at the IEA Congress in 1988 in Sydney (Imada *et al.*, 1988). The third ODAM symposium in Kyoto hosted a session and workshop devoted to participatory ergonomics.

Since 1984 there has been a steadily growing interest in employing participatory methods for implementing ergonomics. The term is becoming more familiar among ergonomists around the world. More importantly, it is consistent with cross-cultural participatory paradigms.

This book first introduces the concept of participatory ergonomics, presents its rationale and overviews the tools used in different participatory applications. The second half of the book uses case studies to demonstrate how participation can extend the utility of ergonomics as an applied discipline.

We owe our thanks to the contributing authors for their outstanding work, their patience and all the staff who made their manuscripts possible. We owe special thanks to Many Ann Leasure and Dixie Imada for their extraordinary efforts in editing and coordinating our work. This was a truly participatory endeavour.

K. Noro
A. S. Imada

References

Imada, A., Noro, K. and Nagamachi, M., 1986, A workshop at the Second ODAM Symposium, August, Vancouver.

Imada, A., Brown, O., Noro, K., Nagamachi, M., Kogi, K., Liker, J. K. and Lifshitz, Y., 1988, A panel discussion on participatory ergonomics at the IEA Congress, Sydney.

Noro, K., 1984, A workshop 1984 International Conference on Occupational Ergonomics, May, Toronto.

Noro, K. and Imada, A., 1984, A workshop on Participatory Approach of Ergonomics, Hawaii.

Noro, K. and Kogi, K., 1985, Invitation to participatory ergonomics, *Anzen* **36**, 53–60.

Sell, R., Noro, K., Imada, A. S., Martensson, L. K. and Kogi, K., 1985, A panel discussion at the IEA Congress in Bournemouth, 1985.

Contributors

L. Chapman,
University of Wisconsin, Madison,
Wisconsin, USA.

S. Dray,
Honeywell Inc., MN26-4201, Honeywell
Plaza, Minneapolis, MN 55408, USA.

A. S. Imada,
Human Factors Department, Institute of
Safety and Systems Management,
University of Southern California, Los
Angeles, CA 90089-0021 USA.

B. S. Joseph,
Corporate Ergonomist, Ford Motor
Company, 105 Central Laboratory,
15000 Century Drive, Dearborn,
Michigan 48120, USA.

K. Kogi,
Occupational Safety and Health Branch,
International Labour Office, 1211 Geneva
22, Switzerland.

J. K. Liker,
Center for Ergonomics, Industrial and
Operations Engineering, University of
Michigan, IOE Building, 1205 Beal
Avenue, Ann Arbor, Michigan 48109,
USA.

K. Miezio,
University of Wisconsin, Madison,
Wisconsin, USA.

M. Nagamachi,
Department of Industrial and Systems
Engineering, College of Engineering,
Hiroshima University, Shitami Saijo-
cho, Higashi Hiroshima 724, Japan.

K. Noro,
School of Human Sciences, Waseda
University, Mikajima 2-579-15,
Tokorozawa, Saitama, Japan.

O. Östberg,
Testing and Research Laboratories, Swedish
Telecommunications Administration, 123
86 Farst, Sweden.

R. G. Rawling,
Principal Consultant Work Environment,
Health and Safety Department, State
Electricity Commission of Victoria, Box
2765Y, GPO Melbourne, Victoria 3001,
Australia.

M. M. Robertson,
Human Factors Department, Institute of
Safety and Systems Management,
University of Southern California, Los
Angeles, CA 90089-0021, USA.

S. S. Ulin,
Center for Ergonomics, Industrial and
   Operations Engineering, University of
   Michigan, Ann Arbor, Michigan 48109,
   USA.

J. R. Wilson,
Institute for Occupational Ergonomics,
   Department of Production Engineering
   and Production Management, University
   of Nottingham, University Park,
   Nottingham NG7 2RD, UK.

K. J. Zink,
University of Kaiserslautern, Kurt-
   Schumacher-Strasse 26, Postfach 3049,
   6750 Kaiserslautern, Germany.

Part I
CONCEPTS, METHODS AND PHILOSOPHY

Chapter 1
Concepts, Methods and People

K. Noro

Concept of participatory ergonomics

The field of ergonomics extends well beyond scientific inquiry. It addresses people and the technology that connects them to their work. The breadth and knowledge required to improve human–machine working relationships make it difficult to confine ergonomic activities to any one organizational or academic field. Ergonomists must work together with nonexperts on a company-wide basis. This procedure is called *participatory ergonomics*. Participatory ergonomics is a new technology for disseminating ergonomic information.

Advantages of participatory ergonomics

The concept of 'fusion' must be understood and implemented for participatory ergonomics to be effective. Fusion refers to a technique for assuring necessary persuasion or cooperation between the expert and nonexpert or for finding a tradeoff between values produced by different solutions and drawing one conclusion.

Advantages of participatory ergonomics are essential for attracting people from different fields. While organizations may cite more advantages, from our perspective there are at least three advantages to employing participation.

1. Efficient utilization and integration of people and information are critical from product development to marketing phases. People are important resources and should be valued by their companies. However, these valuable resources are divided by regions, functions, departments and other organizational boundaries. For example, the personnel and design departments are interested in different types of people. Human factors handled by different departments, such as design, product liability (PL), productivity, reliability and worker health, can be treated on a common basis by the participatory approach. For instance, it will be no less advantageous to share the human resource data of the personnel department and the human capabilities of the development department.

3

At the 3rd ODAM Symposium held in Kyoto in July 1990, participants said that participatory ergonomics is one of the new information technologies in ergonomics. Great attention should be paid to this assertion. The following section features quality-control-circle activities but one should not come to a hasty conclusion that participatory ergonomics is intended only for factory workers. Participatory ergonomics is a new medium, in which everybody can participate, for exchanging information on ergonomics.

2. Ingenious utilization of human information, skill and experience; artificial intelligence (AI) and knowledge engineering fuzzy set theory (Noro, 1990) and their utilization by ergonomics are expected to grow rapidly and are the focus of interest of many people. Apart from academic levels, it is important to build a framework that utilizes the knowledge, skill and experience abundant in the workplace. Participatory ergonomics literally exists for this purpose. Those who regard the participatory approach as low in academic level may be surprised to know that this same participatory approach is required to implement advanced technologies.

3. Considering workers' opinions: the increase in the number of workers with higher education, mental jobs, women workers and worker mobility require that more thought be given to worker participation. Participatory ergonomics can play a major role in meeting these changes. Through participatory ergonomics we can expect some added benefits:

   (i) the designer can receive feedback on what the user thinks about the product;

   (ii) the manager can uncover improvements without inspecting the entire plant; and

   (iii) the worker can make his/her superior aware of his/her interest in his/her job and his/her willingness to work; s/he can also utilize these methods for making improvements in his private life at home.

The following must be recognized to establish the corporate environment in which participatory ergonomics is to be implemented. *Who* has implemented the ergonomics is as important as *what* has been accomplished by ergonomics. It is especially important to note that Japanese companies encourage participatory ergonomic activities in the context of QC circles. Such a system utilizes the intellectual assets that are shared within and between companies. Japan's economic success is greatly owed to a network or a close information-sharing system in Japan. Examples are intimate cooperative relations among companies as represented by Toyota's 'kanban' system, cooperative relations between the government and private sectors, information-exchange systems within corporate groups, and information sharing between management and labour in company unions.

The system of using kanban has been used in conjunction with a just-in-time manufacturing system. Kanban is a Japanese word which literally translated means 'visible record'. These are based on long-standing relations and constitute a close-knit information-sharing system. This is one of the strengths of Japanese companies. This same strength can also be applied to participatory ergonomics.

It is important to ensure the security of corporate secrets. This is probably true in every country. At technical committee meetings researchers from companies in different fields say things that boldly touch upon company secrets. Study groups of this type have flourished in Japan because the participants can learn and exchange valuable information. This is one example of a potential system for sharing intellectual assets. Experts can participate as group members or advisers. They reinforce the idea that who participates is as important as what ergonomics achieves.

Participatory ergonomics as a company-wide activity

Comparing the spread of ergonomics and quality control will help you understand the need for participatory ergonomics. The introduction of quality control was clearly aimed at the spread of scientific management as one means of industrial development in Japan. Quality control made progress peculiar to Japan thereafter. What quality control means for Japanese companies is aptly indicated by the words of Ishikawa (1969): 'In Japan we consider quality control as a revolution in the ideas of management control'. Quality control in Japan is no longer used exclusively by specialists as a scientific technique. It has the significance of a company-wide movement, embracing top management, machine operators and salesmen as well as outside contractors. This company-wide quality control is called total quality control (TQC). TQC is a password for employees of many companies in Japan. Another password of long standing is 'participating approach of quality control'. These quality-control activities have brought product quality to very high levels.

Unlike the quality control that was introduced from Europe and the USA in the 1950s, ergonomics was introduced to Japan through voluntary action of scientists. This brought about a large difference in the subsequent development in Japanese ergonomics and quality control. Japanese companies are more enthusiastic about quality control than they are about ergonomics. The concepts and principles of ergonomics should be helpful to machine designers in the workplace as well as machine operators, housewives, and many other users. Ergonomics is not for a handful of specialists. The participatory approach of involving nonexperts as well as experts is important for the rapid progress of ergonomics. This participatory approach should be followed by all workers on a company-wide basis. Attention has been given to activities by small groups of workers. However, these activities by small groups will not prove effective unless all affected parties (i.e. designers to salesmen) participate in the activities. Two examples of such activities are presented below. One involves the activities of white-collar workers in the computer industry, and the other involves the activities of skilled workers in the steel industry.

Quality-control circles have promoted many activities for sharing information within companies. There are signs that such activities are beginning to develop in ergonomics as well. In recent years, interest in ergonomics has been growing among computer manufacturers in Japan. These firms hold in-house ergonomic lectures, where their engineers can learn the latest ergonomic information. This type of education is accelerating the introduction of ergonomics into the computer industry.

In 1986, industrial designers and others from companies in different fields inaugurated the Ergo-Electronics Technical Committee to exchange information. This committee was established to make products and environments easier for people to use. At their meetings, the committee members exchange knowledge, know-how and data bases for them to perform ergonomic product design. The Ergo-Electronics Technical Committee is composed of 30 corporate ergonomists. Its secretariat is located in the Japan Technology Transfer

Association. The members meet every two months and have a retreat once a year. The Ergo-Electronics Technical Committee produced a video entitled 'Designing Ergonomics into Your Products' to introduce ergonomic products developed by its members. This video was presented in the national program at the 1988 Human Factors Society Annual Meeting. In 1987, eight computer manufacturers organized a technical committee for exchanging ergonomic information under the auspices of the Japan Electronic Industry Development Association.

Another unmistakable sign of the active introduction of ergonomics by Japanese companies is reflected in the themes selected by quality-control circles for their activities. Many of the problems recently analysed by small groups of workers, including quality-control circles, are concerned with ergonomics. One-third of 313 company-commended small-group improvements at the Yawata Works of Nippon Steel Corporation were about ergonomics (Noro, 1988). This report may mark the beginning of the application of ergonomics by workers themselves and may add a new page to the history of ergonomics in Japan. One of the reports presented at a Kyushu area quality-control circle meeting in 1983 was on fault tree analysis by workers (Kishikawa, 1983). The use of such a highly advanced ergonomic technique may cease to be unusual in the near future.

Implementing ergonomic and small-group methods

Small-group activities

At present small groups, mainly quality-control circles, are utilized widely in numerous companies in Japan. Sandholm (1983) and other researchers point out that quality-control circles formed by workers in factories have contributed to the quality control of Japanese manufactured products. Many of the examples given in this chapter show that small-group activities also contribute to productivity improvement. This contribution is too large to be ignored and without it Japanese factories and offices would not have reached today's levels of productivity. For details of the activities of quality-control-circles see JUSE (1987). The history and objective of small-group activities are summarized below.

Objective of small-group activities

Small-group activities are intended to provide rewarding work and self-fulfillment through work. More concretely, groups achieve these goals by developing capability at the workplace and exercising creativity.

History of small-group activities

Small-group activities in Japan started with the zero defect (ZD) drive introduced from the USA in 1965. The ZD drive coexisted with quality-control-circle activities, and spread through the workplace. The workers who participated formed small groups at their respective workplaces. At the end of May 1990, 308 574 quality-control circles were registered at the Quality-Control Circle Headquarters and totalled 2 418 870 members. Figure 1.1 shows the years when small-group activities were started at 433 Japanese

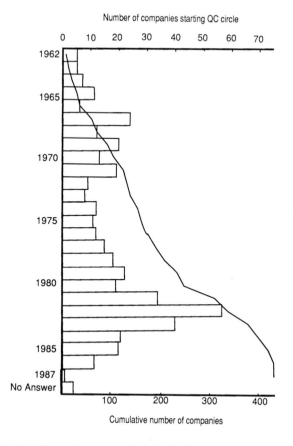

Figure 1.1. Number of companies starting quality-control-circle activities between 1962 and 1987.

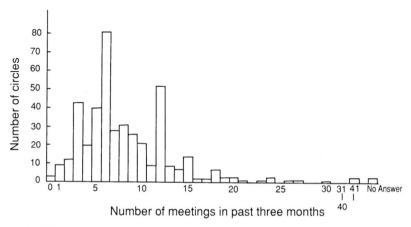

Figure 1.2. The frequency of voluntary quality-control-circle meetings at the respective companies over a 3-month period.

companies (surveyed in January to February 1987) (JUSE, 1987). Figure 1.2 shows the frequency of voluntary meetings of quality-control circles at the respective companies.

Ergonomic improvements included in suggestions for improvements made by small groups to companies and commendation system

Many companies have established a system for implementing suggestions from small groups. Table 1.1 summarizes the replies of 666 establishments in government agencies and private companies to the suggestion survey conducted by the Japan HR Association and the Japan Suggestion Activity Association in fiscal 1989. A total of 60 343 937 suggestions were made at these establishments. Suggestions are evaluated by the company. Workers who have made valuable suggestions are commended. While American and European companies tend to commend individuals for their suggestions, Japanese companies commend employees for the results of their suggestions. The Japanese suggestion system does not commend workers when good results are obtained from their suggestions, but after their suggestions are translated into standard practices. Under this system, good working methods are not held by individual workers as their personal know-how, but are accumulated as the property of their company.

Table 1.1.   Number of suggestions made per member.

Number of suggestions	Number of companies
1 or less per year	35
1 or more per year	49
1 or more per month	62
1 or more per week	3
1 or more per day	1

According to a survey done by the Yawata Works of the Nippon Steel Corporation and Noro (1988), ergonomic studies were included in suggestions for improvements made by small groups there during 1983. During 1983 there were 8829 general suggestions related to manufacturing and development. Less than 5 per cent (313) of these suggestions won special recognition. One-third of these prize-winning suggestions can be classified as ergonomic suggestions. The 109 improvements suggested can be further classified by content as shown in table 1.2.

One example (Noro, 1988) of a type (a) improvement (table 1.2) is a signboard installed to notify crane operators of operating errors. This helped to raise the percentage of crane

Table 1.2.   Classification of 109 improvements according to content (Noro, 1988).

Type of improvement	Number
a Facilitate work or lessen work-load	65
b Follow human characteristics	28
c Combination of the above two	16

operators who passed practical tests from 79 per cent to 90 per cent. Another example of this type of improvement is the wear protection of bends in a powder transport pipeline. Mechanical failures due to the wear of the pipeline bends were reduced to such a degree as to virtually eliminate the use of manual labour to correct such failures. An example (Noro, 1988) of a type (b) improvement is the use of parallel rays from tungsten halogen lamps instead of the previously used fluorescent lamps to reveal scale defects which were otherwise overlooked. An example (Noro, 1988) of a type (c) improvement is the prediction of noninjury accident and resultant change in the running route of an unmanned cargo train line at a steelworks. Another typical example (Kobayashi, 1989) of a type (c) improvement concerns a female worker who performed a task which had been formerly assigned to a male worker. Kobayashi (1989), the only female member in her quality-control circle, investigated tasks conventionally claimed to be too strenuous or dirty for female workers and proposed that a robot arm should be fixed with a torque wrench so that the operation could be performed by female workers. The number of type (c) improvements has been rising recently. As demonstrated by the four cases described below, an effective utilization of ergonomic principles is one of the reasons for the increase in the number of type (c) improvements reported. Mori (1989), a member of the Dolphin Quality-Control Circle, conducted a visual experiment to enhance the legibility of address symbols attached to completed automobiles so that completed automobiles could be readily found for shipment from their depot, and devised more legible address symbols according to the experimental results. Nishiobino (1990) in the Third SGC Quality-Control Circle arranged several keyboard input modes and made possible mode selection by means of push-buttons to reduce keyboard input mistakes.

Murakami (1990), in the Murakami Quality-Control Circle, observed that footstools of the same height were used to inspect the painted condition of the roofs of passenger cars on her painting line and found that the footstools made it difficult to inspect the passenger car roofs and forced the inspectors to assume constrained posture on the footstools. She proposed the introduction of an automatic lifter, and her proposal was adopted.

Selected themes

Table 1.3 shows the five themes most frequently selected by small groups from 1970 to 1979 (as revealed by a survey of 157 factories by the Nikkam Kogyo Shimbun in 1982) and the five themes most frequently selected by quality-control circles at 383 establishments surveyed in January and February 1987 (JUSE, 1987). Ergonomics is helpful for each of these themes.

Table 1.3. Themes selected by small groups and quality-control circles.

Order	1970	1976	1979	1987
1st	Quality	Quality	Cost	Efficiency
2nd	Cost	Cost	Quality	Cost
3rd	Efficiency	Efficiency	Equipment	Quality
4th	Equipment	Safety	Control	Control
5th	Control	Mistake	Safety	Equipment

The five major themes selected by quality-control circles as found by the 1987 survey (JUSE, 1987) are not greatly different from those selected by small groups as shown in table 1.3. However, in the 1987 survey, safety ranked eleventh. This finding means that safety is already assured at the establishments surveyed and is not as great a concern as in the past. Improvement in customer service and increase in sales appeared among the themes in the 1987 survey (JUSE, 1987).

Ergonomic tools

Techniques used by small groups

Among the techniques used in the cases of small-group activities reported in the January and December 1989 issues of the *QC Circle Journal* are the cause-and-effect diagram, Pareto diagram, histogram, scatter diagram, control chart, various graphs, check sheet, and stratification. When these techniques are compared with those revealed by the surveys made from 1970 to 1979, the only technique added is various graphs. The graphs include the so-called matrix and radar charts. The tools used in today's small-group activities are more diversified than those used in the 1970s.

Each technique is easy to use and represents results graphically so that they can be easily understood. Ergonomists and industrial doctors should develop and provide techniques that can be used by small groups to utilize the knowledge of ergonomics at their workplaces to a higher degree.

Using the expertise acquired through these studies

The quality-control circle 'My Road III' (Fujikawa, 1982) of the Autobody Section in Toyota's Tsutsumi Factory, comprises four workers engaged in the autobody welding process. A total of 183 Unimate welding robots are involved in this process. The robots frequently stopped in the middle of operation. The quality-control-circle members found relay contact failure as one of the causes of this problem. This discovery led members to want to know more about the mechanism of the Unimate. They obtained this information from related sections within the factory and from the manufacturer. The workers used this expertise to clarify two other mechanical causes of the problem.

'0.1-s Operation'

The quality-control circle 'Spin' (Suzuki, 1982) at Nippondenso, consists of 10 workers engaged in the fabrication and installation of automotive exhaust gas cleaners. The quality-control circle took the elimination of unreasonableness, unevenness and wastefulness from their work as its objective. The objective was a commonplace one, but the approach the circle members took to achieve the objective was not common. The method the circle adopted is referred to as the '0.1-s operation'. The members commonly understood the length of 0.1 s as the time they took to bend one finger, as shown in figure 1.3.

The circle members studied time-and-motion analysis techniques and video-taped their work. They played back the video-tape recording and reviewed their work procedure. This

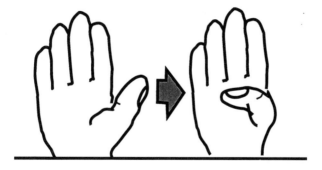

Figure 1.3. Schematic illustration of 0.1 s: the group members commonly understood the length of 0.1 s as being the time they took to bend one finger.

Work area   Motion analysis

Video analysis

Figure 1.4. Analysis of present work procedure as illustrated by quality-control-circle members.

Table 1.4. *Number of improvements made in the '0.1-s operation'.*

Item	Measure	Number
Eliminate unreasonableness	Standardize hand position and orientation	245
Eliminate unevenness	Prevent mistakes when installing and removing parts	358
Eliminate wastefulness	Shorten distance over which parts must be removed	116

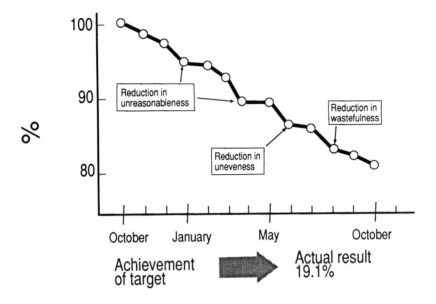

Figure 1.5.   Change with time in achievement of man-hour reduction goal from October to the same month of the following year. The vertical axis indicates the man-hour reduction to be achieved with respect to the present man-hours, which are put at 100 %. Since the actual data are confidential, they are permitted to be published in this format. The actions implemented in the period concerned are described along the curve.

study process was illustrated by them as shown in figure 1.4. The video-tape showed the slight mistakes they made in their procedure due to the difficulty of the work involved. They continued their effort to correct mistakes by improving the tools and apparatus. As a result, the improvements listed in table 1.4 were made. The change in the number of failures during the period concerned is shown in figure 1.5.

Some ergonomists may grimace when they learn that 0.1 s is the time it takes to bend one finger. It should be understood, however, that this idea came from factory workers' experiences. Ergonomic methods for measuring and analysing the motion time are technically correct, but are too difficult to be used by factory workers. In other words, the ergonomist should recognize that the ergonomic methods do not produce any profits if they are unused because they are too difficult. While they may be more scientific than the yardstick devised by the factory workers, they may be less useful.

Discovering problems that require ergonomic solutions

Small-group members do not specialize in ergonomics, nor are they educated in ergonomics. Despite this fact, they often perform activities that can be classified as being concerned with ergonomics. Here, 6 of the 35 reports presented at a Kyushu area quality-control-circle meeting in October 1983 are discussed; particular emphasis is given to the processes that quality-control circles used to identify their specific problem.

Table 1.5.   Steps commonly taken by all small groups.

Step No.	Description
1	Select theme
2	Set goal
3	Grasp present situation and analyse factors
4	Identify problem
5	Develop and improve measures to solve problem
6	Confirm effect of measure taken

Almost all small groups are educated beforehand to follow similar steps (see table 1.5), each of which is described below.

Step 1: Selecting a theme

In the report presented by the LF circle (Suzuki, 1982), latent accidents were selected as the theme. Another theme is the use of new techniques to enhance activity levels of the circle members. The Tanishi circle (Uesugi, 1983) selected the reduction in electric arc welding noninjury accident experienced by workers as their theme ('hazard' being used as an approximate equivalent to the Japanese word '*hiyari hatto*' that means the mental condition into which one falls when one is faced with danger). The report stated that the theme met the company's slogan: 'Be sure to deliver safe and good products at low cost'.

Step 2: Establishing a goal

All quality-control circles have quantifiable goals to be attained. In their reports, the LF and Tanishi circles both mention improvement in productivity and safety as their goals. The goals established by six quality-control circles are presented in table 1.6. The table indicates that the two most common goals are reduction in time and improvement in safety. It goes without saying that human factors are extremely important for each goal to be achieved.

Table 1.6.   Goals set by six quality-control circles.

Eliminate latent accidents and halve working time (Uesugi, 1983)
Eliminate noninjury accidents and halve working time (Kikuchi, 1983)
Increase number of articles inspected per person and accordingly
    reduce personnel by 25% (Iwamizu, 1983)
Raise productivity by 8% (Fujii, 1983)
Eliminate noninjury accidents (Kishikawa, 1983)
Shorten maintenance time by 30% (Nagata, 1983)

Step 3: Understanding the situation and analysing factors

The reports of the six quality-control circles mentioned above tell us that small-group activities can be characterized by two features. The first is that the circle members all follow

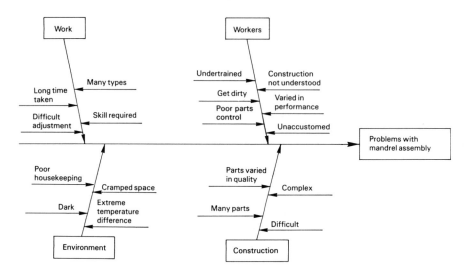

Figure 1.6.   Typical cause-and-effect diagram used in small-group activities.

the same process: they observe their workplace to understand the present situation; recognize one common goal; review the workplace with the goal in mind; and discover new facts from the workplace. The second feature is that they use various techniques in this step. One motivation for these activities is to try new techniques which they have not used before. In the report of the Threader circle (Nagata, 1983) the cause-and-effect diagram, which is the tool most frequently used in small-group activities, was employed to identify the factors in question. The cause-and-effect diagram prepared by the quality-control circle is shown in figure 1.6. The other circles also included practically the same factors (working methods, workers and environment) in their cause-and-effect diagrams.

Pareto diagrams are also used to analyse waiting time (Iwamizu, 1983) and to analyse working time (Kikuchi, 1983). Many circles try to predict noninjury accident at unsafe points in processes before accidents actually occur (Kishikawa, 1983; Kikuchi, 1983). In this activity, the circle members record the hazardous conditions ('*hiyari hatto*') which they experience in their work. This type of activity is called 'hazard prediction'.

One quality-control circle (Fujii, 1983) observed the operating condition of each machine, while another (Kishikawa, 1983) conducted fault-tree analysis of noninjury accident in work.

Table 1.7.   *Classification of causes (human errors) of accidents in the material tension test at high temperature.*

Lack of confirmation
Unfamiliarity
Misoperation
Failure to observe standards
No knowledge of standards
Inadequacy of standards

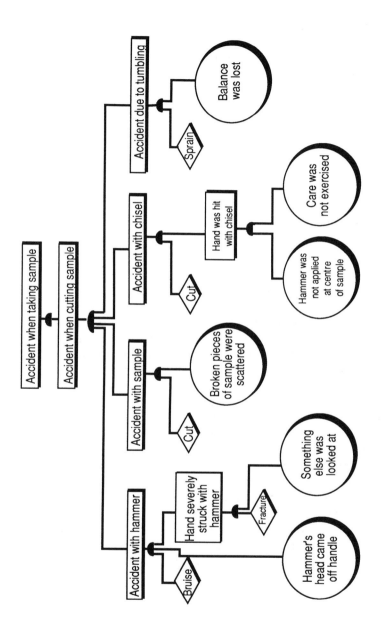

Figure 1.7. A fault tree constructed by the 'material testing' circle.

Figure 1.7 shows a fault tree constructed by the 'material testing' circle. The material-testing circle (Kishikawa, 1983) classified the cause of accidents as shown in table 1.7 for the operations in which many errors were made in the past. They recognized these errors as human errors and established corrective measures.

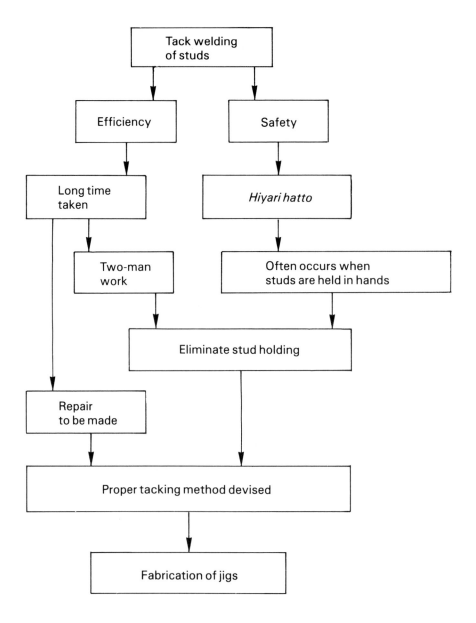

Figure 1.8.   Flow-chart for the pursuit of working efficiency and safety at the same time.

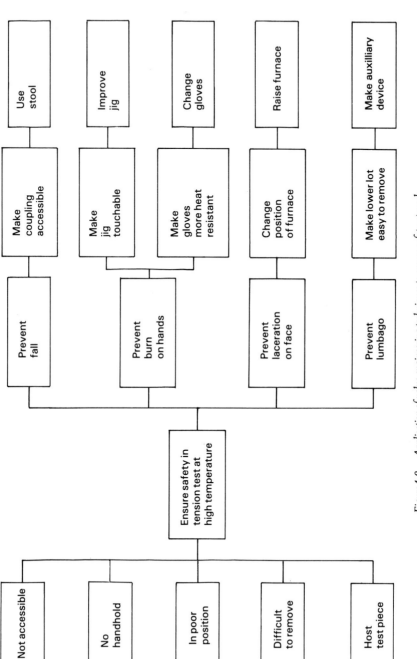

Figure 1.9.   Application of value-engineering techniques to ensure safety at work.

Step 4: Identify problem

Step 4 involves identifying the causes of problems. The Tanishi circle (Kikuchi, 1983) found that tack welding of studs took the longest time of all operations in their work. Circle members asked themselves why the tack-welding operation took longer than the other operations. They used a cause-and-effect diagram, and combined the working-time analysis and safety analysis of this operation in one flow-chart (see figure 1.8). Figure 1.8 indicates that the circle members treated the problems of efficiency and safety with equal importance.

Step 5: Develop and improve measures to solve the problem

The material-testing circle applied value-engineering techniques to ensure safety in their material tension testing work (Kishikawa, 1983). The application of value-engineering techniques is illustrated in figure 1.9.

Step 6: Confirm effect of measure taken

A diagram similar to figure 1.11 shows a typical action carried out by a small group to confirm the effect of the measure taken. This diagram allows us to confirm at a glance the magnitude of the effect obtained and the contributing factors.

Five ergonomic viewpoints

The worker knows his workplace better than anyone else and this is the advantage which a worker has over any specialists when analysing the workplace and solving the problems associated with it. However, because a worker is accustomed to the workplace, he is likely to

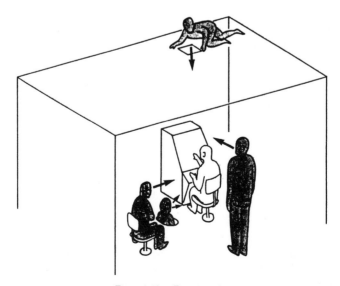

Figure 1.10. Five viewpoints.

Table 1.8. *Five viewpoints from which to inspect work.*

From directly above
From eye level in standing posture
From the same level as desk
From diagonally below worker's eye level
From foot level

overlook improvements. This is the disadvantage of observing one's own workplace. Table 1.8 and figure 1.10 provide unconventional viewpoints for observing one's own workplace.

Five ergonomic tools

Ergonomic methods that can be used at the workplace must be simple and carefully thought out. Textbooks cannot be brought near the machines. The ergonomic methods must be simple enough to be remembered. Manuals or guides should be small enough to be carried in a pocket. Five techniques (tools) meet these criteria and are summarized in table 1.9. We need to instruct small-group members in the ergonomic methods listed in table 1.9. They must recognize that the methods for understanding human behaviour differ from typical small-group techniques.

Table 1.9. *Five ergonomic tools recommended for use by workers at workplace.*

Technique	Concept	Remarks
Time analysis	Speed	Pareto chart can be used
Man–machine chart (Iwamizu, 1983)		Logo of '0.1-s operation' (Suzuki, 1982)
Fault-tree analysis	Mistake	
Noninjury accident (*hiyari hatto*)	Feeling of worker	
Easy-to-draw chart	Dimensions and postures of people and objects	For example, posture weight points of Toyota Body (JDAE, 1982)
Force	Ready chart	

Guidelines for applying participatory ergonomic techniques

Restrictions on use of ergonomics at the workplace

Participatory ergonomics should be promoted in education at the workplace and, if possible, by the foreman briefly before the start of the day's work. Those in charge of ergonomic education and research should consider developing ergonomic methods. As already described, ergonomic methods are important driving forces for the small-group activities of workers. Only a few methods can be used at the workplace due to the constraints shown in table 1.10.

Table 1.10. Barriers to applying ergonomic methods in industry.

1. Measuring instrumentation for ergonomic research, common in universities and research institutes, is not available at many companies and factories
2. If the factory is equipped with such instruments, measurements are influenced by a variety of artifacts
3. Management does not understand the gains to be made through the application of ergonomics
4. Company staff and factory workers do not understand the application of ergonomics
5. There are few corporate ergonomists who have mastered the latest measuring techniques

For the widespread application and acceptance of ergonomics in industry, the ergonomist must prepare methods that enable measurements to be made in actual production fields (such as factories) and help everyone concerned understand the results obtained. Such methods must conform to the policy given in table 1.11.

Whatever theme is selected, the starting point is to understand the workplace. This knowledge of the present situation promotes the discovery of a theme. Once a theme is selected, the small group works toward meeting that goal using simple methods. For example, a single method may comprise five ergonomic viewpoints and five ergonomic tools.

Table 1.11. Methods that ergonomics should provide for industry.

Simple but selected methods: measurements are on site, using a few methods selected from among those studied

People responsible for ergonomics

Japanese companies have no universal departments or sections responsible for ergonomics. People with ergonomics responsibilities are those listed in table 1.12. These people are specialists who, in addition to their original tasks, perform inspection or improvement as necessary by using ergonomics. The results of a survey done to determine those responsible for ergonomics at five Japanese companies are shown in table 1.13 (*Nikkei Mechanical*, 1986).

In periods of high economic growth, Japanese factories were automated by introducing robotics and the flexible manufacturing system (FMS) as priority technologies for enhancing

Table 1.12. People in charge of ergonomics at Japanese companies.

Engineers
  Design engineers
  Production engineers
  Process-control engineers
Safety and health managers
Industrial doctors
Industrial health nurses
Small-group activity members

Table 1.13. Results of a survey done to determine the people in charge or ergonomics at workplaces in Japan (Nikkei Mechanical, *1986*).

Person responsible	Company
Manager of hygiene in occupational safety and health department	Toyota Motor
Factory job redesigner	Sakai factory of Daikin Industries
Factory design room and four workers	Machinery division of Niigata Engineering
General manager of production (stress alleviation) engineering department	Okuma Machinery Works
Manager of welfare section in personnel department	Okuma Machinery Works
Representative of labour union	Okuma Machinery Works
Industrial physician	Okuma Machinery Works
General administration department in charge of labour and personnel	Tatebayashi factory of Fujitsu

productivity and efficiency. As the economy slowed, some companies started to utilize ergonomics in their operations. *Nikkei Mechanical* (1986) gave the following examples of these moves:

1. utilization of ergonomics to support the Toyota production system;
2. development of evaluation measures to lessen workload, such as posture weight point;
3. development of a portable computer for inspecting posture weight point and upper limb points;
4. job redesign for elderly workers at Daikin Industries;
5. application of ergonomic techniques to tyre manufacturing at Yokohama Rubber;
6. improvement in working environment at Graphtec; and
7. control of FMS by workers at Niigata Engineering.

The greatest impact of participatory ergonomic activities is in production. The production manager is a key player in the participatory ergonomics team. This person's main interest is to meet desired production levels while maintaining or improving health. Experience from quality-control-circle activities suggests that the production manager needs knowledge of ergonomics and methods that can be utilized in the field.

Cost/benefit evaluation of small-group activities

Europeans and Americans often ask questions about the cost/benefit ratio of small-group activities. Corporate managers would not suspend small-group activities because of an unfavourable cost/benefit ratio. This is not the sole criterion for continuing small-group activities. However, both managers and workers show great interest in the cost/benefit ratio of small-group activities. Most small groups report the benefits of their activities to their companies in written summaries. Their description of benefits may be classified as: qualitative effects; reduction in working time or number of *hiyarri hatto* (noninjury accident); and the amount of benefits.

The CADokko circle (Miwa, 1983) cited the following qualitative effects: the circle was awarded a grand prize at a quality-control-circle meeting of the factory and the circle members shook hands with the factory manager. And a circle member's mother-in-law understood quality-control-circle activities.

Some small groups cite an increased sense of responsibility and reduction of stereotypes as further benefits arising from small-group activities. The reduction in working time and number of *hiyarri hatto* (noninjury accident) is the most frequently cited benefit of small-group activities. For instance, a small group of workers (Uchimaru, 1983) at a workplace where several types of engine cylinder blocks were machined undertook activities to reduce the set-up time. The set-up time after the corrective measures were implemented was cut by half (see figure 1.11). This type of illustration is used by a great number of small groups. Many small groups also calculate benefits in terms of money.

Strategies for introducing and implementing participatory ergonomics into organizations

This section describes the methods of persuading companies to implement participatory ergonomics and the efforts required to help workers understand participatory ergonomics.

Explanation of need for strategies for implementing ergonomics in organizations

Ergonomists may be the only people who believe ergonomics should be included as part of corporate activities. Many examples of improvement and development have been presented at many meetings, including the IEA Congress. These improvements and developments have not brought about true benefits to companies as often as ergonomists might wish to think. Behind each success story lies a process for examining whether ergonomics can coexist with required productivity levels. Participatory ergonomics (defined here as ergonomics introduced via small-group activities semivoluntarily performed by workers with the support of their companies) may not be immediately accepted by companies. Participatory ergonomics will not succeed without the understanding of employers and managers. The conditions described in this section can be understood by workers at Japanese companies but may not be understood by others. In Sweden, for instance, the working environment of workers is protected by ordinances and companies are legally bound to improve harmful working conditions. In contrast, if Japanese workers want an improvement in their working environment, they must submit a proposal form that shows how this improvement will benefit the company. This proposal is then reviewed and approved by the manager. This unlegislated approach may lead one to conclude that Japanese companies may be considered inferior to their European counterparts. However, as a result of participation, the working environments in Japanese factories are among the best in the world.

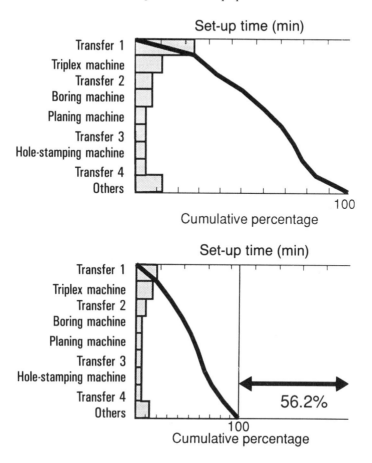

Figure 1.11. Illustration of benefit obtained through activities to reduce set-up time: (a) before and (b) after implementation of corrective measures. Similar graphical illustrations are used by many small groups.

Menu of persuasive methods

In general, the main interest of corporate management is to improve productivity. It is not easy to show how ergonomics can contribute to improving productivity. There are many employers and managers who believe that the introduction of ergonomics decreases, rather than increases, productivity and does not bring about any profits to the companies.

Methods of persuading companies to adopt participatory ergonomics and helping the workers to understand participatory ergonomics are described below. Some ergonomic experts may think that these persuasive methods are too down-to-earth and cumbersome for their research area. This may be true. In fact, nonexperts in the field cannot obtain a working environment they wish to have unless they overcome the complexity of problem solving. The persuasive methods are described below. The following examples are concrete applications of the concept of fusion discussed earlier.

First persuasive method

Participatory ergonomics is an important part of the suggestion system.

Step 1. Participatory ergonomics is explained so that both company management and workers can readily understand it and use it to obtain a comfortable working environment and working conditions.

Step 2. The workers make proposals for improving their working conditions to the management together with proposals for improving and maintaining productivity.

Step 3. Participatory ergonomics is explained so that both management and workers can readily understand how it creates a comfortable working environment and working conditions.

The effort expended to maintain productivity refers to the level of comfort; i.e. workers are not exposed to too great a work-load. In other words, full consideration must be given to the ease with which the workers can perform their assigned tasks.

Second persuasive method

The following benefits can be obtained from the improvement of an uncomfortable work environment.

Step 1. To create an easy-to-work environment, eliminate the difficulty of working. An easy-to-work environment is one in which equipment is easy to operate, production efficiency is kept consistently high for a long period of time, and safety is assured.

Step 2. The difficulty of working means a heavy work-load. The heavy work-load may give rise to intense fatigue, unpleasantness or high stress.

Step 3. If a difficult-to-work environment is improved, productivity can be increased, product quality can be stabilized, the work-load can be alleviated, safety can be increased, or the skills of middle-aged and older workers can be put to effective use.

Third persuasive method

In contrast to the two methods described above, the following explanation is effective for a manager who is not satisfied unless presented with specific figures or graphical representations. It will be most effective if the manager is presented with data on work efficiency in his own operation. The workers make proposals for improving their working conditions to the management together with those for improving and maintaining productivity.

Figure 1.12 shows a sign that was posted in the aisle of a certain factory. Figure 1.12 shows that the x component must be maintained for the company to improve its productivity. The term 'maintenance' is a concept that can be discussed by using time as a yardstick. Figure 1.12 explains productivity alone. Figure 1.13 schematically illustrates the relationship between productivity and ergonomics (comfort is used as an alternative characteristic here).

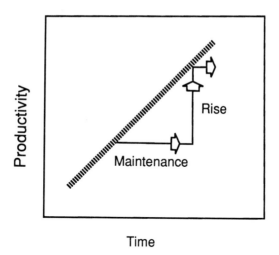

Figure 1.12. A sign posted in the aisle of a certain factory. The productivity gain is accomplished by the maintenance of the x component and the rise in the y component.

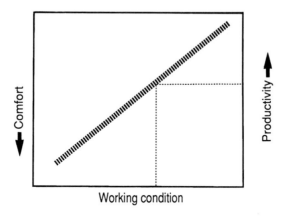

Figure 1.13. Schematic illustration of the relationship between productivity and ergonomics (comfort is used as an alternative characteristic here). Straight line for determining working condition with productivity and comfort as opposing measures.

Assume that the goal of the workers is opposed to that of the company management. The working condition is determined at some point on the straight line. The best solution cannot be obtained from figure 1.13 alone. When one looks at the change in productivity with time (as shown in figures 1.14 and 1.15), one finds which condition to select.

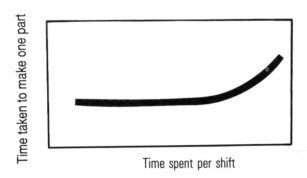

Figure 1.14.   Work cannot be maintained under the condition of maximum productivity.

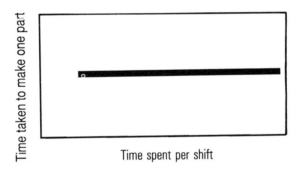

Figure 1.15.   Work can be maintained under the condition of good comfort.

Example

Workers in a machine factory organized a group to find ways of reducing their fatigue when they worked bending at the waist. The waist bending of the worker shown in figure 1.16 was measured to be 15°. It is presumed that if the height of the circular machine is increased by a few centimeters the worker can reduce the frequency of bending at the waist and perform his/her task more comfortably over a longer period. This reason was not good enough for managers to permit the necessary change to be made. Already cognizant of this possibility, the group then prepared figure 1.17. There should be no such waist bending, as shown in figure 1.16, but, in this case, the optimum viewing distance cannot be obtained. As a result of discussion, the workers concluded that the optimum waist-bending range over which the optimum viewing distance and comfort requirements could both be satisfied was 8° to 12°. It is already known that under the condition of constant machine floor elevation, some waist bending is better for the viewing distance, and hence work efficiency, than no waist bending.

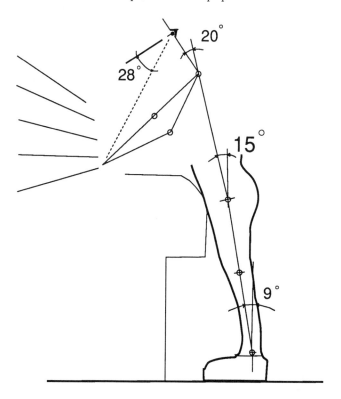

Figure 1.16.   Illustration of the relationship between worker A and his/her machine. The waist angle is 15°.

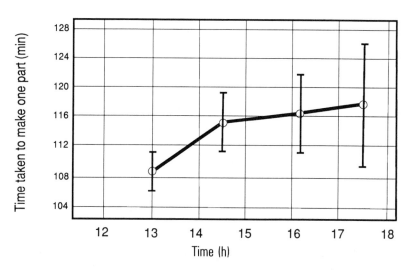

Figure 1.17.   The time taken to make one part at the end of work in the afternoon was about 10 per cent longer than at the start of work and the variability in the number of parts made increased at the end of work as compared with the start of work.

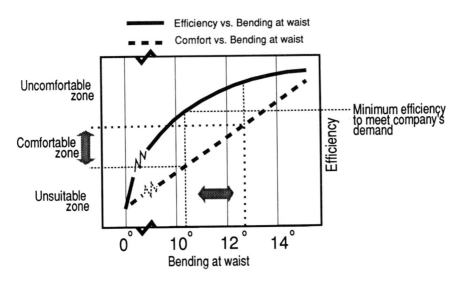

Figure 1.18. The relationship between the three conditions: bending at waist, ease of working, and work efficiency. If the actual values measured at a factory are plotted, the working conditions that meet both the ergonomic and production requirements the factory can be established.

To summarize, both the workers and managers are satisfied when waist bending is limited to between 10° and 12°. The vertical axis of figure 1.18 is zoned according to this finding. This diagram is helpful in persuading both the company's management and labour that their requirements can be met. Figure 1.18 shows the three concepts (efficiency, bending at the waist and comfort) at the same time. The scale of comfort was prepared based on the subjective survey of the workers themselves in the group. This type of treatment is rarely observed in the research area but is observed daily in the field. The efforts made to persuade the company's management of the advantages of participatory ergonomics and to help workers understand participatory ergonomics may be 'onerous' efforts that the ergonomic experts do not make. These efforts must be expended to change the minds of employers and managers who believe that ergonomics reduces productivity and brings no profits to their companies. The ergonomic experts should support these efforts more willingly. Examples of the fusion technique are described above.

References

Fujii, H. (VIP circle), 1983, Improvement in productivity by increase in machine stroke rate, *1367th Proceedings of QC Circle Kyushu Area Meeting*, Kitakyushu: Kitakyushu Division of QC Circle (Secretariat: Irie Kousan Co.), pp. 69–72.

Fujikawa, K. (QC Circle 'My Road III'), 1982, Reduction in number of Unimate failures, *QC Circle Reports JUSE*, pp. 72–3.

Ishikawa, K., 1969, Education and training of quality control in Japanese industry, *Reports of Statistical Application Research, JUSE*, **16**, 21–40.

Iwamizu, E. (Deming circle), 1983, Improvement in productivity of toilet stool inspection, *1367th Proceedings of QC Circle Kyushu Area Meeting*, Kitakyushu: Kitakyushu Division of QC Circle (Secretariat: Irie Kousan Co.), pp. 69–72.

JDAE (Job Development Association for the Elderly), 1982, Research Report on Job Redesign.

JUSE, 1987, *Third Survey of Actual State of QC Activities*, Tokyo: JUSE Report.

Kikuchi, M. (Tanishi circle), 1983, Improvement in stud attaching method, *1367th Proceedings of QC Circle Kyushu Area Meeting*, Kitakyushu: Kitakyushu Division of QC Circle (Secretariat: Irie Kousan Co.), pp. 37–40.

Kishikawa, T. (Material testing circle), 1983, Standardization of safe work in high-temperature material tension testing, *1367th Proceedings of QC Circle Kyushu Area Meeting*, Kitakyushu: Kitakyushu Division of QC Circle (Secretariat: Irie Kousan Co.), pp. 96–9.

Kobayashi, H., 1989, I was also able to make robots, *Proceedings of 2420th QC Circle Meeting, Fukuoka*, pp. 60–1.

Miwa, Y., 1983, Preparation of error-free original drawings, CADokko Circle, Yaskawa Electric Manufacturing Co. Ltd, *1367th Proceedings of QC Circle Kyushu Area Meeting*, Kitakyushu: Kitakyushu Division of QC Circle (Secretariat: Irie Kousan Co.), pp. 143—6.

Mori, H., 1989, You can more easily find completed automobiles for shipment, *Proceedings of 2380th QC Circle Meeting, Nagaoka*, pp. 156–7.

Murakami, M., 1990, Improvement in as-painted surface appearance of passenger cars 'Charade', *Proceedings of 2480th QC Circle Meeting, Kanazawa*, pp. 98–9.

Nagata, M. (Threader circle), 1983, Improvement in mandrel rod, *1367th Proceedings of QC Circle Kyushu Area Meeting*, Kitakyushu: Kitakyushu Division of QC Circle (Secretariat: Irie Kousan Co.), pp. 123–6.

Nikkei Mechanical, 8 September 1986, pp. 42–7.

Nishiobino, M., 1990, Prevention of frequent, short robot line stops, *Proceedings of 2480th QC Circle Meeting, Kanazawa*, pp. 78–9.

Noro, K., 1988, Participatory ergonomics, *Japanese Journal of Ergonomics*, **24**, 5–10.

Noro, K., 1990, Participatory ergonomics: concept, advantages and Japanese cases, in Noro, K. and Brown Jr, O. (Eds), *Human Factors in Organizational Design and Management*, Amsterdam: North Holland, pp. 83–6.

Sandholm, L., 1983, Japanese quality circles — remedy for the West's quality problem?, *Quality Problems*, **XVI**, 20–3.

Suzuki, K. (QC Circle 'Spin'), 1982, 0.1-sec. Operation We Tackled, *QC Circle Reports*, pp. 178–179.

Uchimaru, K., 1983, Activities to halve setup time, Nissan Motor Co. Ltd, *1367th Proceedings of QC Circle Kyushu Area Meeting*, Kitakyushu: Kitakyushu Division of QC Circle (Secretariat: Irie Kousan Co.), pp. 33–6.

Uesugi, A. (LF Circle), 1983, Standardization and education of safe work when taking gas samples for analysis, *1367th Proceedings of QC Circle Kyushu Area Meeting*, Kitakyushu: Kitakyushu Division of QC Circle (Secretariat: Irie Kousan Co.), pp. 21–4.

Yamada, K., 1990, Report on actual improvement and suggestion activities, *Soi to Kufu*, No. 109, pp. 1–5. In Japanese.

Chapter 2
The Rationale and Tools of Participatory Ergonomics

A. S. Imada

Introduction

Over the past few years, people have begun to realize the vital role that ergonomics can play in increasing productivity, quality, safety, and overall organizational effectiveness. By optimizing the match between people and machine systems, ergonomists have improved systems and human performance. The success of ergonomics is evident in advertising, product liability litigation, and product design. Furthermore, ergonomics has extended beyond the simple matches between humans and machines (knobs and dials, controls and displays) to cognitive areas such as software design, decision-making, and decision facilitation. That first level of knobs and dials can be thought of as a first generation of ergonomics. The cognitive area, when dealing with mental processes, can be thought of as a second generation of ergonomics. A third level incorporates these human–machine matches into a larger organizational context. Hendrick (1985) calls this third generation 'macroergonomics'. This third-generation view holds that ergonomics, like any other technology, cannot exist independently of organizational and management considerations. Therefore, when introducing ergonomics we need to consider the organizational culture, management system, communication patterns, rewards, and a host of other variables that support ergonomics as a technology.

Participatory ergonomics provides one perspective in macroergonomics. This term was first coined by Noro and Imada in 1984 (Noro and Imada, 1984). The major premise of this concept is that the ergonomics is limited by the degree to which people are involved in introducing this technology. In its simplest terms, participatory ergonomics requires that end-users (the beneficiaries of ergonomics) be vitally involved in developing and implementing the technology (Imada *et al.*, 1986; Noro *et al.*, 1986).

There are at least three compelling reasons which argue for involving people in the development of the ergonomic technology.

1. Ergonomics in and of itself is an intuitive science. In many cases it simply provides names and labels for ideas, principles, or practices that workers are already using. In

one sense it legitimizes the ideas and experiences that workers have accumulated in the process of doing their jobs.

2. Ownership in ideas enhances the likelihood of implementing ergonomics successfully. People are more likely to support projects for which they feel ownership. In the long run, this has implications for a more involved and dedicated work-force committed to problem solving.

3. End-user participation developing technology creates a flexible problem-solving tool. That is, if people implement the technology, they will be able to modify it to solve future problems.

Therefore, when the business environment or process requirements change, the users then mould the technology to fit the problem. Under the traditional scenario an ergonomic consultant is hired to solve the problem. When the problem emerges again or is modified, the solution needs to be reintroduced with the consultant. While this may make economic sense in the short term, it has negative effects in the long run. This is true for two reasons. First, problem solving is done on a case-by-case basis, and is not generalizable to other situations. Second, the message that it conveys to the worker is 'You are not capable of solving problems. Problem solving should be left to experts. You are not an expert. Do what you are told'. Participation, however, recognizes the worker as a valuable resource for solving problems. Recognizing this expertise enhances workers' self-esteem.

The chapters in this book demonstrate how participation at all levels enhances the impact that ergonomics can have on work life. These interventions cover a range of settings. They occur at national, organizational, and shop-floor levels. The interventions span the nature of the work being performed, from a labour-intensive manufacturing operation to information-processing organizations; from high technology computer organizations to third world developing countries. All these case studies clearly demonstrate that the value of participatory ergonomics lies not in the methodology or the tools of participation, but in enabling people to participate in developing, designing, and utilizing ergonomics to improve their work. While these case studies differ in levels of technology, the skill level of the people, organizational level, and cultures, they demonstrate one point simply but effectively, i.e. the effect that ergonomics can and does have depends largely on people's involvement and understanding. In some cases, the technology or the tools used to solve the problem are not as important as the effect it has on people's mind set. If people are actively thinking about their work, safety, productivity, comfort, and the ease of work they can improve the overall quality of work life (QWL). For example, in Chapter 9, the tools that were designed by the DELTA group were not used as originally intended. However, using the tools did allow people in the organization to accomplish what they needed to do, within their own organizational culture. In this case, the DELTA group and the DELTA framework are tools for achieving an end, and not an end in themselves. Like innovators who introduce new technologies or managers who design systems, ergonomists can get caught up in the tool itself rather than the goals they are trying to achieve.

An important characteristic of participatory ergonomic techniques is that they enable people to understand and apply ergonomics to their work. These tools need to be simple and directed at the audience and problems that they are intended to solve. Whenever possible, the authors have included examples of the tools that they used in implementing the solutions.

When using these tools, remember that just as these authors have modified the tools to fit their particular needs, so you too must make modifications to these tools to fit your needs and the people that you serve through ergonomics. The tools and methods used are simple and robust enough to be modified and transferred to different scenarios. Several examples come to mind. First, Wilson's example (Chapter 5) of involving librarians and cashiers in redesigning their work stations, involves simple methodologies that almost appear 'unprofessional'. However, this was ideal for the design decision groups (DDGs) who needed to make design changes quickly and easily while still keeping a good perspective of anthropometrics and work processes.

The chapters by Nagamachi (Chapter 7) and Zink (Chapter 8) show how we can use quality circle methods to introduce ergonomics. By using these tools, both authors have been able to solve physiological problems as well as complex psycho-social problems in manufacturing operations in Japan and Germany. Despite differences in language and methods, the ideas remain the same. The ideas are simple, intuitive, and make the user aware of the causes of the problems; they empower the user to analyse, solve, and overcome these problems themselves.

The third and perhaps most stringent example of meeting user needs is presented in the chapter by Kogi (Chapter 4). Participation is carried out in developing countries where not only is the technology foreign to the people, but the technology exists in a hostile environment. Kogi's use of appropriately simple intervention and feedback mechanisms is effective in meeting these workers' needs. The guidelines that Kogi provides for simplicity, take-home value, and immediate cost/benefit are certainly generalizable to other situations.

The case studies described in this book represent important success stories in developing a third generation of ergonomics. They establish a foundation for solving a diverse set of problems using ergonomics. If ergonomics is to be successful in solving problems, we must *empower* and *enable* people to apply the technology.

The argument for participation

We can begin by asking: Why should people participate in implementing technologies? Our everyday lives are surrounded by increasing technological issues and the question may really be translated to: Can workers absorb and understand the technology that surrounds them? People living in free societies are often asked to make decisions about technologies that affect their lives. Can people understand the technologies for which they are making decisions? Similarly, employees who are asked to participate are requested to make decisions about their work and how the work is to be done.

At a broader level, the question can be rephrased as: Can our societies and organizations continue if people do not understand the technology associated with their work? If people do not feel ownership in the systems they work in, can we expect our societies to survive? If not, we are likely to see a greater degree of worker alienation and job dissatisfaction. Several years ago a study undertaken at the Carnegie-Mellon University revealed that 25 per cent of the working robots in the USA were not operational because workers: (a) did not know how to make them work; or (b) did not want to make them work (Byham, 1983). This type of worker alienation has no place in a highly competitive international market-place. Unless organizations and societies reject advancing technologies, which is not really possible, two scenarios are plausible. In the first, all technical issues and technical matters are 'managed' by an elite class of technical bureaucrats. The actions of these bureaucrats will be largely sheltered and screened by the very technology they manage. These technical caretakers need little input from people who are affected by the technology or people who use the technology. Organizations and societies will be governed by technological tyrants. In the second scenario, end-users are actively involved in the system, understand the system, and make informed decisions about using the new technology to make work easier, less dangerous and generally more satisfying. However, to do this people need to understand the advantages and disadvantages of technology, how the technologies are being used, and how to enhance QWL.

The chapters in this book demonstrate how organizations and social systems can act to keep people from being ruled by technological tyranny. Nagamachi's example (Chapter 7) of the Kubota Automation Key Man System demonstrates that workers can introduce flexible manufacturing systems (FMS), robotics design, and ergonomics in unmanned systems. Zink (Chapter 8) suggests that worker participation can reduce such maladies as gastro-intestinal illness and general stress. From an organizational standpoint these examples demonstrate that people can understand and solve the problems which they face.

Arguably the two greatest social changes in current American society are: the civil-rights movement, and the women's movement. Both movements have affected the way we interact with people, the way we think about people, and the opportunities we provide for people. What is most notable about these social movements is that they were never professionally organized, legislated nor mandated. Rather, both were started at a grass-roots level. People sensed an injustice and became involved. This kind of participation has tremendous value for creating social change. Similarly, ergonomics can create social change by improving QWL and improving organizational outcomes through participation. Like the two large-scale social movements, we need to involve people, and not rely on ergonomists to solve problems.

An argument for participation can be made from four perspectives: historical, socio-psyˈ ɪological, organizational, and technological.

Historical perspective

It may be helpful for ergonomists and users of ergonomics to view the profession from an historian's perspective. Boorstin, librarian for the US Congress, argues that ergonomists and other professionals may be a threat to industrial democracy (see Boorstin, 1986). He argues that the beginnings of professionalism were the first deviations from amateurism. Interestingly, the word amateur comes from the Latin word *amator* (lover; someone who does something for the love of it). This person pursues goals not for money, fame, glory, prestige nor promotion; but because this person simply *loves* doing it. This person cannot help doing what he or she does. Just as an aristocrat governs by virtue of birthright, technocrats govern because of their knowledge. The members of this technological elite are not necessarily those who want to or know how to make systems work. It is the amateurs — those who love their work, love their job — that truly understand the problems, concerns, and suffering of work.

Interestingly, when the word 'profession' first emerged in the English language, it related to the vow taken by the clergy. By the 16th century professionals in other vocations used the word to describe 'a professed knowledge of some department of learning, or science used in the application of affairs of others' (Boorstin, 1986, p. 22). After several centuries these professions began to create tight monopolies within schools and colleges. About 1840, the words 'scientist' and 'artist' entered the language. All over Europe scientists and artists began to achieve respectability. What is perhaps most interesting in Boorstin's analysis is the premise of the professional fallacy. The fallacy is that the profession exists for the sake of the professionals it represents. That universities exist not for teaching, but for professors; that laws exist for lawyers; and that journals exist for professionals who can publish. According to Boorstin, 'beneath the professional fallacy lies the confident axiom that the customer is not competent to judge. Any professional — whether a brain surgeon or a plumber — must command our faith in his expertise' (Boorstin, 1986, p. 23).

One by-product of the professional fallacy is that the term 'amateur' now carries a negative meaning. It implies that the work is either unprofessional or crude. Likewise, many academic disciplines designed to solve everyday problems have now become scientific. In so doing these disciplines have become more complicated and removed from the potential beneficiaries they are intended to serve. Therefore, to be less than scientific is to be unprofessional. We can fall victim to the professional fallacy.

In sharp contrast, participatory ergonomics proposes to make the end-user a vital part of this scientific methodology. For example, Kogi's principle of building on local practices may defy the scientific rigour of experimental studies. However, it is extremely well suited for the users of this technology. Using what he calls 'learning by doing', Kogi has successfully implemented 80% of the solutions within a 30-day period in developing countries. Here the user is part of the solution rather than a passive recipient of a professional's technology.

Like the great social movements of our time, ergonomics can make a contribution by actively involving people in its implementation. Like an enduring art form, ergonomics must have meaning to the people that use its principles and methods. Moreover, these principles and methods must be adaptable to specific situations. This can be achieved by avoiding the professional fallacy. That is, by involving people in its practice the profession can exist for the benefit of the user.

The solutions to complex problems and human suffering can be disarmingly simple. There is great elegance in this simplicity. For millenia people have been shaping and modifying tools to fit the needs of specific tasks. This is evidence that people have been practising ergonomics throughout history. These tools — whether primitive arrow heads, knives, chisels, or computers — reflect a trial-and-error process that makes work easier, safer, and more efficient. Conceptually this is no different from what we are trying to achieve today. The differences lie at the levels of technology and the integration of different technologies that are brought to bear on today's problems.

To take its place in human history as a tool for creating change, ergonomics must incorporate the user. People will be unwilling to apply a technology from which they realize no benefit. People will be less willing to become submissive recipients of a technology. Ergonomics will be an effective technology if we avoid falling victim to the 'professional fallacy'.

Socio-psychological perspective

Perhaps because we live in a technologically advanced society, we have come to expect certain characteristics in our work. We know that people derive their identity and meaningfulness from work. However, as workers have changed so have those identities and meanings. Today's better trained and better educated workers have expectations greater than basic pay, benefits and a safe place to work. These expectations include participating in meaningful decisions. People value self-expression through work and look for opportunities to be recognized. Today's worker may want the opportunity to work with interesting people and participate in decisions about how work is done (see, for example, Yankelovich, 1979). This is perhaps appropriate since the new technologies require people to have more control over their work and to become problem solvers (Lawler, 1986).

While technology in organizations creates opportunities and expectations for improved QWL it also created jobs that fall short of these expectations. Many of the jobs in today's high-technology organizations belong to what Toffler (1980) calls the 'second wave'. These jobs can be characterized by regimentation, coordination, standardization, specialization, synchronization, concentration, maximization, and centralization. Jobs in second-wave systems do not allow people to use discretion or to make decisions. They have short task cycles which require high repetition of the same activity. They specify in detail how the activity should be carried out and they require little interaction among people. We have known for a long time that highly routine jobs, especially ones that are machine paced, cause stress. A great deal of this stress can be traced to the loss of control over timing and discretion. The likely outcomes of these job characteristics are job dissatisfaction, turnover, absenteeism, and possibly sabotage (Hackman and Oldham, 1980).

Motivation and satisfaction

For many years 'humanistic' writers (e.g. Argyris, 1957; McGregor, 1960; Likert, 1961; Lawler, 1986) have advocated the benefits of getting people to participate in organizational decisions. Early studies (Lewin, 1943; Coch and French, 1948); demonstrated the power of involving groups in discussions and decisions. We believe that there are psychological benefits to allowing people to participate. Participation allows people to set goals. There is evidence that setting goals has a positive effect on performance, particularly difficult goals (Latham and Yukl, 1975; Zander, 1979; Latham and Saari, 1979) and work satisfaction (Locke and Schweiger, 1979). The information-sharing aspect of participation also makes participation a powerful tool. Researchers have found that this information sharing can motivate people involved in solving problems (Latham and Saari, 1979; Locke and Schweiger, 1979; Locke *et al.*, 1981; Bartlem and Locke, 1982).

There is now an emerging body of evidence to suggest that participation can have positive effects. For example, studies indicate that participating in quality circles can affect productivity (Donovan and Van Horn, 1980; Tortorich *et al.*, 1981; Benscotter, 1983; Srinivison, 1983; Wolfe, 1985), quality (Hunt 1981; Jenkins and Shimada, 1983), absenteeism (Hunt, 1981; Tortorich *et al.*, 1981; Krigsman, 1984; Mohrman and Novelli, 1985; Wolfe, 1985), grievance rates (Hunt, 1981; Totorich *et al.*, 1981), job satisfaction (Zahra, 1982; Jenkins and Shimada, 1983; Rafaeli, 1985; Shores, 1985; Wolfe, 1985), and organizational commitments (Hunt, 1981; Horn, 1982; Benjamin, 1983; Griffin and Wayne, 1984; Platten, 1984; Seybolt and Johnson, 1984, 1985; Wolfe, 1985).

Finally, there is evidence that participation at work may have the greatest benefit for unionized and minority women. Waldron and Jacobs (1988) have reported that this workplace participation had the greatest health benefit for unmarried and minority women. They postulated that this may be due to the social support participation which these workers do not get outside of work.

'Small wins'

Another psychological dynamic that may operate in participation has been described by Weick (1984) as 'small wins'. Weick argues that the inability of social science to solve large-scale problems is due to problem definition. Large-scale social problems defy human problem-solving. Overcoming such massive problems as hunger, coronary heart disease, crime, pollution in cities, and traffic congestion, is beyond the scope of human capability. These problems, Weick argues, can be more easily solved if the problems are scaled to human proportions. Problems must be recast into less arousing proportions, that help people solve the problem. If we create situations that allow people to work on the problem and solution directly, we may be able to stabilize the degree to which people are aroused. The solution is a series of 'small wins'. Small wins are concrete, complete, implemented outcomes of average importance. The wins by themselves may seem unimportant. However, a series of these small wins can show a pattern of progress that may arouse others to act, and prevent opponents from resisting. The beauty of the small win is that once it occurs the next solvable problem becomes more visible. This may occur because others begin to see the same problem and begin synergistic problem-solving.

Small wins may also provide ways for people to learn new skills and learn more about their jobs. Small wins are preferable over large wins, because they are more structurally sound. They are more stable if they are built one on the other. Peabody (1971) cites Saul Alinsky's criteria for building community organizations. Actions need to be specific, realizable, and meaningful. If people are able to work on something that is concrete they can experience visible success. They develop confidence in their actions which can translate into an excitement and optimism that becomes contagious. Small wins help achieve these goals.

Perhaps the lesson to be learned from Weick's analysis is that when people are confronted with such massive problems as 'productivity', 'safety', 'profits', and 'market share' they are overwhelmed. There is little that one person can do about these large-scale organizational problems. However, redefining the problem in terms of individual actions may motivate people to react and respond. Furthermore, the employee may be motivated if he/she can define what actions should be taken. If the employee realizes success in this 'small win' the next improvement is more motivating. At least four interesting psychological dynamics may operate with small wins and participation.

Creating change

Many organizational problems require large-scale changes. People are uncomfortable with such changes. If the change can be defined at moderate levels, it causes less arousal and less resistance. An example may be an alcoholic or someone who is trying to quite smoking. Both realize that this will require a life-style change. However, life-style change is too large a concept to think about when you want a cigarette or drink. The success of Alcoholics

Anonymous may lie in its ability to cast the problem in small wins. Although the goal is to change a life-style, the first level of analysis is to get through the next hour. That is a small win. Likewise it is too difficult for a smoker to think about living without cigarettes forever. However, if one can think about making it through some period of time that is much more manageable, and reward oneself, the goal becomes much more palatable. These small wins are not only easier to accept, they are more pleasurable to experience. It is more motivating for people to experience a series of small victories than to work for a long period without being rewarded.

The research on attitude change suggests that people are more likely to comply with a series of small requests than a large-scale change (Friedman and Fraser, 1966). Furthermore, rather than creating large differences, it is probably more useful to create just noticeable differences (JND) so that people are more receptive to changes of moderate intensities rather than overwhelming intensities (Davis, 1971).

Stress

An important facet of stress is the perception that one does not have the abilities or skills to deal with stressors in the environment. Therefore, when confronted with large-scale problems people feel that there is nothing they can do. We might reduce stress if we can redefine these problems to a level that people can act. For example, it can be stressful for people to live or work in an unpredictable environment where they feel there is very little they can do to act upon that environment. However, if through their actions they are able to affect that environment, they become the locus of control. Creating a comprehensible concrete gain, however small, may change perceptions about control. This may in turn cause people to act with confidence and forcefulness in dealing with controllable world. Taken together, these dynamics may make large-scale problems less stressful and improvements clearer.

Bounded rationality

The third psychological dynamic may be the notion of 'bounded rationality' (March and Simon, 1958; Perrow, 1981). Bounded rationality predicts that people can only deal with limited amounts of information at one time. Consequently, people take perceptual shortcuts in processing information (Kahneman *et al.*, 1982; Kiesler and Sproull, 1982; Miller and Cantor, 1982). Small wins may be able to minimize the distortion people experience through these heuristics. Small wins allow people to visualize, manage, and monitor the kind of information necessary to solve real problems. Problems such as 'Our company cannot withstand international competition' exceed limits of bounded rationality. However, redefining the problem to 'What can *I* do to help our company be more competitive?' allows people to solve problems rather than to mull over uncontrollable aspects of their world.

Many of the participatory ergonomic tools in this book represent methods of achieving small wins. Check lists, radar charts, Pareto diagrams, organizational assessment tools, and quality circles are means for helping people define and achieve small wins. The elegance of ergonomics is that it complies nicely with Weick's definition of 'small wins'. We can create a series of concrete, complete outcomes of moderate importance through ergonomic changes. It allows people to think about their problems in concrete ways.

The challenges for achieving these small wins are two-fold. First, we need to motivate people to get involved in achieving these small wins. These ergonomic tools serve as invitations for people to solve larger problems. Second, a series of small wins must lead to visible progress. However, defining the next small win is not always easy. They may appear unrelated because they may occur in a non-linear fashion. Despite these challenges, the potential benefits of participatory ergonomics through small wins are considerable. There are sound psychological reasons for assuming that allowing people to deal in simple, concrete, and bounded gains will result in solving much larger problems.

Self-determination

Participation in determining work and work outcomes seems to be consistent with current trends in self-determination. Perhaps because of the self-help experiences in the 1970s, people began to believe that they can control their environment and their lives. The current view of health and fitness is a good example. There is a large-scale movement in American society toward health and fitness through diet and exercise. The logic is that people are responsible for what they put into and what they do with their body. Prior to this, it was generally believed that addictions were the result of forces beyond the control of those afflicted. Behaviour was tolerated. Therefore, those public figures exhibiting unacceptable behaviour were often forgiven because they were viewed as victims of their environment or of their jobs. More recently, however, people have begun to believe that we are responsible for our own actions. The public is no longer tolerant of abuse which is deemed to be within one's own control. This emphasis toward self-control and self-determination places the responsibility of controlling behaviour with individuals. This social trend is consistent with why more workers want control of their work. Participatory ergonomics simply provides the tools for allowing people to exercise that control.

Organizational perspective

A number of writers (e.g. Naisbitt, 1982; Lawler, 1986) have been predicting the demise of centralized bureaucratic organizations. The inability of these traditional designs to meet societal and organizational needs has made networking and participation critical. The pyramidal structure so common in describing organizational design and reporting relationships is no longer relevant because of the need to exchange information quickly. While networking and participation take more time, they are more effective in dealing with complex problems. For example, one common criticism of Japanese decision by consensus is that it takes a long time to make a decision. The advantage, however, is that once consensus has been reached, decisions can be implemented and changed quickly because all affected persons know their role, and how the decision affects their unit, their department, or their jobs.

What caused this need for organizations to rely on networking rather than hierarchical structures and layers? While there is no single reason, we can point to a number of factors that necessitate organizations becoming flatter, more participative, and more network oriented.

The nature of business

Rapidly changing technology and customer needs have created shorter product life cycles, longer production times, and shorter lead times. To respond to these changes, organizations have had to become less bureaucratic and more flexible. Response time is critical. The ability to put a product on a market first has important market value. The risk of taking the product to market first involves quality and user-friendliness. If the product is not of high quality and user-friendly, the competitive advantage of being first is lost. Therefore, implementing change quickly is critical to organizational success. Involving people at different levels of the organization can improve speed.

The international market-place

Critical factors in international competition include labour costs, productivity, and regulatory and governmental influence. In many cases, companies in industrially developed countries are competing with wages only a fraction of their labour costs. Industrially developed countries are competing with one another for the same market share. The difference between success and failure in these competitive markets may depend on worker participation. For example, Imada (1982) has reported that quality-control-circle activities in a Japanese manufacturing company produced 18 000 implemented suggestions per month. This means that there are nearly a quarter of a million suggestions implemented on an annual basis for improving quality, productivity, and safety. To offset this advantage competing organizations must be able to realize cost savings or production gains through similar participation programs, capital improvements, or increased technology. At best, the capital improvement and added technology strategies will have marginal returns in the long term. Moreover, peoples' problem-solving ability enables the same skills to be recycled. It is clear that the participative strategy is more effective in the long term.

New products and technology

As technology becomes more specialized, organizations require more specialized labour. People with increased abilities and specific skills now deal with the technical problems. Centralization, maximization, coordination, synchronization, and other 'second-wave' methods of management are not appropriate for these skilled employees. These industrial engineering concepts, once the backbone of industrial production, are no longer as applicable to workers whose linkages are more fluid, lateral, and free flowing. Being able to network across levels in an organization is as important as vertical support. A good example is given in Chapter 9. New technologies need to be introduced with business, technological, and organizational perspectives in mind. All three perspectives are critical to successful change. This dependence on networking makes specialized workers more important; perhaps as important as the capital assets that the technology requires. Without networking the system is not able to function as designed.

Human input to reliability

While humans may play smaller roles in automated systems, their roles become more critical. System reliability can only be maintained through human participation. For example, as technology increases, the experience of a maintenance person in dealing with a particular situation decreases. Consequently, this person has less experience of dealing with particular malfunctions. This lack of experience, in turn, leads to higher reliance on technical resources such as manuals, guides, and computer decision aids. Unless the equipment is completely self-checking, maintenance personnel become more critical. Therefore, while there may be fewer personnel, their functions become more critical.

The solution most often cited is more training. Weick (1987) has pointed out that this is insufficient, and perhaps detrimental, because it does not increase the workers' skills enough to sense the dangers. Furthermore, training people to maintain high-reliability systems is often not motivating because they were maintaining a dynamic non-event. That is, when a system is operating smoothly there is nothing to watch. The cues that people need to sense are severe aberrations in highly reliable machine systems. We can compensate for this by increasing the collective 'requisite variety' of the group by allowing them to interact and participate.

Workforce

The workforce is now better educated. Lawler quotes US Department of Labor statistics that show a 14 per cent increase in the number of people with high-school diplomas entering the work-force. Lawler also cites a recent Census Bureau report revealing that 86 per cent of today's 25–29 year olds already possess a high-school diploma. In sharp contrast, employees of the 'second wave' may have been more docile, uneducated immigrants looking for good wages, good benefits, and job security. For today's labour force these factors are treated as entitlements. Organizations will have to meet needs for interesting work, work with interesting people, and a voice in deciding how work is to be performed. Participation provides one opportunity.

Given the nature of products, international competition, the increased significance of humans, and work-force changes, participation seems to be a logical answer to improving organizational effectiveness. However, this shift cannot be made simply by changing organizational structure. To improve our current functioning will mean an organic change in the way we do business. Allowing people to participate means shifting information and power throughout the organization. Furthermore, doing business by 'telling' will have to be replaced by 'coordinating'. Giving people decision-making authority and decision-making opportunities also means allowing them greater control, authority, and responsibility. It may mean allowing them to perform work in ways best suited to their needs and specific job needs.

A good example is in telecommuting (see Imada, 1986) where workers are allowed to perform work at home. There are several distinct advantages to telecommuting. First, for the organization, there is no longer the need to house these employees. Given the cost of office space in urban areas, this reduced cost is significant. Second, from the workers' perspective there is no need to fight commuter traffic, greater freedom, more time to spend with their family, and more flexibility. Presumably, the many advantages of telecommuting will mean greater productivity and satisfaction. However, how, when, where, and with whom the work is to be done are no longer under the control of the employer. Employees will have more tactical responsibilities for ensuring that work gets done. This requires greater power sharing.

Technological perspective

Technology and change

Participation can be useful not only in implementing technologies but also in helping organizations cope with the changes that technology creates. Several writers have commented on the nature of change that technology creates (see Toffler, 1980; Maccoby, 1984). First, technology by its very nature helps introduce new technology. This creates change. Toffler points out that not only does technology create change, it also creates change at a faster rate. The other implication is that technology creates competition. Any advantage realized through technology (e.g. making products cheaper and faster) is lost once the competition realizes the technological advantage. For example, if a farmer is able to harvest his crop using a combine, neighbouring farms have one of two alternatives: either incorporate the technology and gain the same advantage or go out of business. Therefore, the second farmer must introduce the same or better technology to survive. The first farmer must justify the capital investment of the machine technology and regain competitive advantage. He does this by introducing more or better technology. This continues until the return on the technology becomes marginal.

These series of relationships between technology, change, and competition can be used to explain why technology has grown at an exponential rate. In his chapter on technological change Toffler (1980) describes how the sum total of inventions in the first 79 generations cannot equal the inventions in the 80th generation. However, this growth cannot continue indefinitely. Our ability to apply new technology is limited by two human factors:

1. peoples' *ability* to change; and
2. peoples' *willingness* to change.

We will discuss the former of these in the next section. However, affecting the willingness to change comes through understanding change. If we can get people to understand change, we can probably affect their willingness to accept change. Participation is an excellent vehicle for both understanding change and creating a willingness to change.

Matching humans to complex technologies

Weick's (1987) analysis or 'requisite variety' provides an interesting framework that has implications for participation. Weick argued that system reliability is dependent on matching human abilities to system requirements. Accident events occur because humans are insufficiently complex to sense and anticipate problems in the system. Systems would be more efficient if there were a better match between system and human complexity. This can be achieved in one of two ways: either make the system less complex or get the humans to be more complex. Since technology does not allow systems to become less complex, we must be able to make humans more complex (Imada, 1987). These are hardly new ideas in ergonomics.

What is new are Weick's notions of how we make humans sufficiently complex to deal with technically complex systems. He offers at least two suggestions. First, group activities, especially among heterogeneous groups, can increase requisite variety. People from different professions or different parts of a factory have different mind sets. This causes them to look for different things when trying to understand a problem. Collectively these people would see more than any individual, thereby increasing the group's requisite variety. This in turn improves system reliability and system functioning. Group activities such as quality circles provide an excellent opportunity for creating this kind of requisite variety.

The second implication of Weick's work may be that we need to integrate humans back into complex systems. In large complex systems, people perform fragmented tasks from which they may not be able to see relationships to the whole. If people do not understand what they are doing, and how it relates to the whole, they may not understand how they undermine system reliability. People understand technologies best when they *are* the technology; or at least part of the technology. Machine-paced work or work that requires people to watch machines does not make people part of the technology. These types of jobs do not allow people to understand how they can create hazards in the environment. However, workers who are part of implementing a technology, understand the technology and are more likely to make contributions. Nagamachi's example (Chapter 7) of Kubota's AKMS demonstrates how people can implement and design robotic systems to meet specific needs. Here, humans are an integral part of the technology. Even if the operator does not interact with people while the machine is operating, interaction and participation in meetings to discuss ergonomic problems can make that person part of that technology. Operators begin to see more clearly their role in making the system more efficient and safer.

Taken together, the historical, socio-psychological, organizational, and technical demands on today's systems require people to be involved in implementing new technologies. Ergonomics is such a technology. These arguments suggest that if ergonomics is to make a real contribution, we need to involve people in implementing these ideas.

Tools for participatory ergonomics

This section provides an overview of the tools used in participatory ergonomics. This coverage is not intended to limit users. Instead it familiarizes users with the tools and stimulates new ideas for other tools. Other tools have been used successfully in implementing ergonomics but are not included here because they are beyond the scope of this book. Finally, these tools are not limited to ergonomics. The fact that they come from other disciplines is evidence that they are robust and generalizable.

Pareto analysis

This analysis is intended to separate the 'significant few' from the 'trivial many'. This analysis is based on Vilfredo Pareto's observation that 80 per cent of the wealth in 19th century Italy was controlled by 20 per cent of the people. This is also known as the 80/20 rule. This analysis was made popular by Juran and Deming and their work in statistical quality control. An example of this analysis can be found in Chapter 7, figure 7.8. An important feature of this tool is that it illustrates very clearly to the user what the major cause of the problem is likely to be. This quantitative analysis prioritizes causes from greatest to least impact in a single graph. The analysis combines a bar graph with a Lorenzian curve, which relates the cumulative effects of different causes to problems. Users are able to grasp relationships among these causes when using this diagram. For example, by placing two Pareto analyses side by side, we are also able to plot the progress we have made through different ergonomic interventions. This positive feedback can be motivating to those solving a problem. Most importantly, the Pareto diagram helps people identify the factors that have the greatest influence on the problem being solved.

Cause-and-effect diagram

This tool, sometimes called a 'fishbone chart', is useful when the problem is undefined or unstructured. It is a good group facilitation technique for helping people identify potential causes of the problem and possible interrelationships. Unlike the Pareto diagram, which is quantitative, this qualitative analysis allows people to restructure their own thinking and agree on potential solutions.

A problem is specified as clearly as possible and entered on the right-hand side of the diagram. The four branches that lead to that problem are called the four Ms. These Ms represent the four potential causes: manpower, machinery, method and material. Each potential cause is classified under one of these four Ms. The listing of ideas is limited to the M being discussed. This procedure helps to structure people's thinking about the particular manpower requirement, for example, that causes the problem. The general form of this tool is shown in figure 2.1. Another example can be found in Chapter 1. Going through this analysis also helps people understand the relationships among causes. Like the Pareto analysis, the cause-and-effect analysis is commonly used in quality circles. The success of this process is dependent upon the cooperation of end-users and related parties. Some guidelines for using the cause-and-effect analysis include the following.

1. Full participation will ensure that no causes are overlooked. All participants should feel free to voice their ideas.
2. Encourage free information exchange: a non-critical free participation atmosphere will help this flow. Write down all ideas under the appropriate category.
3. If causes become too cumbersome, group causes together and analyse these groups separately.
4. Select the most likely cause after all ideas have been aired. Only at this point should ideas be evaluated.
5. Focus on solutions. Avoid discussing how the problem started or who is to blame. Focus on solving the problem.

Quantitative illustrations

There are a number of quantitative illustrations that can be used to give people feedback and to provide an analysis of the problem, and are very useful in analysing and helping people to grasp the problem. These techniques include pie charts, bar graphs, histograms, radar charts (see Chapter 7, figure 7.9), and icons. Whatever the technique, the idea is to provide the quantitative information or feedback in a simple and effective form so that users can solve problems.

Five ergonomic viewpoints

Noro (1984) has suggested the use of five ergonomic viewpoints for workers to analyse their workplace. Noro argued that no one knows the workplace better than the worker. However, the worker may be disadvantaged by the fact that he or she may be accustomed to viewing his or her workplace from a particular vantage. Noro's analysis gets workers to look at their work stations from five viewpoints: above, eye level in a standing position, eye level from a sitting position, diagonally from below the workers' eye level, and from foot level. By providing these vantage points Noro has been able to get people to think about their work and the potential ergonomic causes to their problems in different ways than they are normally accustomed to doing. These new orientations are helpful for people to restructure their thinking to different ergonomic ideas.

Link analysis

This analysis is intended to chart the movement of people, information, work processes, or movements on a control board, by plotting the distances travelled (for more information see Kantowitz and Sorkin (1983)). Imada has used this technology method to allow hotel workers in China to understand why orders were taking so long to be processed in a restaurant. It made the causes of these delays clear to management and workers.

Link analyses are powerful tools for participation because they are intuitive, quantitative and illustrative. They allow people to visualize inefficiencies, redundancies, and wasted energy that contribute to location problems. The output can be the actual distance travelled, the frequency of different movements, or the percentages of movement or interactions among people, places or things.

Check-lists

As a broad category, check-lists are useful for analysing problems, evaluating problems, or implementing new technologies. Check-lists can get people to do things they have already learned, but may not remember otherwise (Geis, 1984). Check-lists are also helpful when people have specific procedures to follow (Swezey, 1987). The way in which the screen-checker system is used in Sweden to allow workers to evaluate VDTs is described in Chapter 10. The ways in which check-lists can be useful in helping to prioritize different elements of the workplace is illustrated in Chapter 4 (see also Kogi, 1985). Chapter 3 shows how check-lists can be useful in evaluating different technologies in an office.

Kogi's work using check-lists in developing countries has been especially effective. Check-lists are effective when users have limited reading skills, when the work environment does not allow for time or space to read or write, or when extensive writing is strongly incompatible with the original culture.

World map

Wilson uses a procedure adapted from O'Brien's (1980, 1981) work on using a 5–10 min exercise to define poor problem space (see Chapter 5, figure 5.2). In this exercise participants are asked to list the kinds of equipment, tools, or objects being worked on in that environment. This is an easy exercise that helps people to think about ergonomics and its implications.

Round-robin questionnaire

Here the participants are given an open-ended statement such as 'The problem with my work station is . . . '. Each statement is presented on a separate sheet of paper. The sheet of paper is passed around and each participant must add an idea that is different from the preceding one. This is done until there are no more new ideas or when every participant has had at least one chance to fill out the questionnaire. Unlike traditional questionnaires, this procedure cuts out redundant answers and stimualtes ideas one on the other. People have an opportunity to express ideas without having to go through a long discussion period.

Layout modelling and mock-ups

It may be helpful to provide a three-dimensional tool for workers to evaluate new work stations or new work environments. Imada and Shoquist (1984) discuss the importance of a model in describing to people what a reconfigured work space would look like. Wilson (Chapter 5) goes a step further by having workers actually construct cardboard models of potential work areas. This use of models and mock-ups is useful in helping people to visualize potential changes in their workplace.

Slides/videos

It is helpful for workers to see their work environment and work processes in slides and videos. Kogi's use of instant (Polaroid) slides is useful in the environments he works in because it provides immediate feedback to even the most remote organizations. Wilson's use of slides to study work processes is also a good example of this tool.

In his current work, Imada advocates the use of video-taping actual employees who can advocate or demonstrate the use of safety procedures. Such a visual aid has two advantages. First, it presents a sequence of actions, spatial orientations, or work processes to be presented in a visual mode. This is more effective than the traditional ways of conducting occupational safety and health training (e.g. lectures or demonstrations by an authority figure). The source of the messages and the environment in which it is filmed are also credible. Second, it allows people to participate in the process. This may be valuable to the learner and the model as well. Finally, it is cost effective because the realism need not be recreated over and over again. The labour costs of case involvement can be reduced greatly.

Key learning points and pictoral analysis

Participatory tools should simplify ergonomic analysis into its subcomponent parts. Learning points for complex measurements can be broken down into specific areas of analysis. Two examples are helpful. First, Liker *et al.* in Chapter 6 present a pictoral man–SAM (statistical analysis measurement). Here the workers evaluate stressors to specific joints or body parts and their displacements. This serves as a good starting point for discussion in solving the problem. Second, Nagamachi's use of learning points is dsesigned to teach people how to lift (see Chapter 7, figures 7.10 and 7.11 and tables 7.3 and 7.4). Two aspects of this approach are worth noting:

1. it breaks the task down into its component parts; and
2. the illustrations are in a form that is very common to the culture's cartoon characters.

Organizational design and management considerations

These tools are starting points for helping us incorporate users into the ergonomic process. Ergonomics should be viewed as processes rather than single outcomes. Just as work continues to evolve, so too, workers must be continually involved in helping that environment change.

The case studies presented in this book are excellent examples of what can happen when people are allowed to participate in implementing ergonomics technology. However, participation in the technology alone is insufficient for creating change. Simply allowing people to participate, or teaching them techniques to solve problems is not an end in itself. To be truly successful in participatory ergonomics, we must *empower* people to make decisions and to implement and evaluate them. This involves a dramatic change in the way that organizations are typically run. Allowing people to use these tools without giving them responsibility or authority simply provides lip service.

To implement participatory ergonomics effectively we may also have to create consistent organizational change. This change may have to be at a governmental level (see Chapter 10) or at the very least, a managerial level (see Chapter 9).

As Imada (1985) has pointed out, participation may be an excellent vehicle for helping people meet their individual needs for meaningful and rewarding work and, at the same time, helping organizations meet needs created by increased competition, technology, and change. What this suggests is that it is insufficient to change the technology. An organizational change is required to realise the full effects of participatory ergonomics. It means a shift in power, organizational design, and the way we have done business in the past. However painful this process may be, if successful, we may begin to tap into what seems to be an unlimited human potential for solving problems.

References

Argyris, C., 1957, *Personality and Organization*, New York: Roe.
Bartlem, C. S. and Locke, E. A., 1981, The Coch and French study: A critique and reinterpretation, *Human Relations*, **34**, 555–6.

Benjamin, E. R., 1983, Participation and the attitude of organizational commitment: a study of quality circles, *Dissertation Abstracts International*, **43**, 3062B.

Benscotter, Jr., G. M., 1983, A study of the effects of training in the use of selected quality circle data collection and analysis tools upon leader/group interaction patterns and a group's ability to extract factual information from a data set, *Dissertation Abstracts International*, **44**, 368A.

Boorstin, D. J., 1986, Democracy's secret virtue, *U.S. News and World Report* **30 Dec. 1985–6 Jan.** 22–5.

Byham, W. C., 1983, The steel-collar worker and the I/O psychologist, *The Industrial/Organizational Psychologist*, **20**, 16–21.

Coch, L. and French, J. R.P., 1948, Overcoming resistance to change, *Human Relations*, **1**, 512–32.

Davis, M. S., 1971, That's interesting: towards a phenomenology of sociology and a sociology of phenomenology, *Philosophy of Social Science*, **1**, 309–44.

Donovan, M. and Van Horn, B., 1980, Quality circle program evaluation, *Transactions of the Second Annual IAQC Conference*, **5**, 96–101.

Fingrett, H., *Heavy Drinking: the Myth of Alcoholism as a Disease.*

Friedman, J. L. and Fraser, S. C., 1966, Compliance without pressure: the foot-in-the-door technique, *Journal of Personality and Social Psychology*, **4**, 195–202.

Geis, G. L., 1984, Checklisting, *Journal of Instructional Development*, **7**, 2–9.

Griffin, R. W. and Wayne, S. J., 1984, A field study of effective and less-effective quality circles, *Proceedings of the Academy of Management*, **44**, 217–21.

Hackman, J. R. and Oldham, G. R., 1980, *Work Redesign*, Reading, Massachusetts: Addison Wesley.

Hendrick, H. W., 1985, Macroergonomics, the third generation of human factors, *Japanese Journal of Ergonomics*, **21**, 248–52.

Horn, L. J., 1982, Effects of quality circles on productivity attitudes of naval air rework facility production employees, *Technical Report 82-120-7*, Washington, DC: Industrial College of the Armed Forces.

Hunt, B., 1981, Measuring results in a quality circles pilot test, *The Quality Circles Journal*, **5**, 26–9.

Imada, A. S., 1982, 'Productivity and worker involvement in the Japanese organization', presented at the 26th Annual Meeting of the Human Factors Society, October, Seattle, Washington.

Imada, A. S., 1985, Participatory ergonomics — its utility, its appeal, and its necessity, in Brown, I. D., Goldsmith, R., Coombs, K. and Sinclair, M. A. (Eds) *Ergonomics International*, pp. 364–6, London: Taylor & Francis.

Imada, A. S., 1986, Implementing system and technological changes through ergonomics and organizational design and management, *Japanese Journal of Ergonomics*, **22**, 22–6.

Imada, A. S., 1987, Managing human and machine system requirements through participation: the need for an integrated organizational culture, in Noro, K. (Ed.) *Occupational Health and Safety in Automation and Robotics*, pp. 387–98, London: Taylor & Francis.

Imada, A. S. and Shoquist, D., 1984, Managing ergonomic and process design changes, in Hendrick, H. W. and Brown, Jr, O. (Eds) *Human Factors in Organizational Design and Management*, pp. 569–73, Amsterdam: Elsevier/North Holland.

Imada, A. S., Noro, K. and Nagamachi, M., 1986, Participatory ergonomics: methods for improving individual and organizational effectiveness, in Brown, Jr, O. and Hendrick, H. W. (Eds) *Human Factors in Organizations and Management*, Vol. II, pp. 403–406, Amsterdam: North Holland.

Jenkins, K. M. and Shimada, J. Y., 1983, 'Effects of quality control circles on worker performance: a field experiment', presented at the Annual Meeting of the Academy of Management, Dallas, Texas.

Kahneman, D., Slovic, P. and Tversky, A. (Eds) 1982, *Judgement under Uncertainty: Heuristics and Biases*, Cambridge, England: Cambridge University Press.

Kantowitz, B. M. and Sorkin, R. D., 1983, *Human Factors*, New York: Wiley.

Kiesler, S. and Sproull, L., 1982, Managerial response to changing environments: perspectives on problem sensing from social cognition, *Administrative Science Quarterly*, **27**, 548–70.

Kogi, K., 1985, *Improving Working Conditions in Small Enterprises in Developing Asia*, Geneva: ILO.

Krigsman, N., 1984, 'Quality control circles and feedback: effects on productivity and absenteeism', unpublished PhD, thesis, Hofstra University.

Latham, G. P. and Saari, L. M., 1979, The effects of holding goal difficulty constant on assigned and participatively set goals, *Academy of Management Journal*, **22**, 163–8.

Latham, G. P. and Yukl, G. A., 1975, Assigned versus participative goal setting with educated and uneducated woodworkers, *Journal of Applied Psychology*, **60**, 299–302.

Lawler, III, E. E., 1986, *High-involvement Management*, San Francisco: Jossey Bass.

Lewin, K., 1943, Forces behind food habits and methods of change, *Bulletin of the National Resource Council*, **108**, 35–65.

Likert, R., 1961, *New Patterns of Management*, New York: Hill.

Locke, E. A. and Schweiger, D. M., 1979, Participation in decision-making: one more look, in Staw, M. (Ed.) *Research in Organizational Behavior*, Vol. 1, pp. 265–340, Greenwich, Connecticut: JAI Press.

Locke, E. A., Shaw, N. K., Saari, L. M. and Latham, G. P., 1981, Goal setting and task performance: 1969–1980, *Psychological Bulletin*, **90**, 125–50.

Maccoby, M., 1984, 'Keynote address', presented at the First International Symposium on Human Factors in Organizational Design and Management, August, Honolulu, Hawaii.

March, J. G. and Simon, H. A., 1958, *Organizations*, New York: John Wiley & Sons.

McGregor, D., 1960, *The Human Side of Enterprise*, New York: McGraw-Hill.

Miller, G. A. and Cantor, N., 1982, Book review of Nisbett and Ross, 'Human Inference', *Social Cognition*, **1**, 83–93.

Mohrman, S. A. and Novelli, Jr., L., 1985, Beyond testimonials: learning from a quality circles programme, *Journal of Occupational Behaviour*, **6**, 93–110.

Naisbitt, J., 1982, *Megatrends*, New York: Warner.

Noro, K., 1984, 'Small group activities in quality control circles in Japanese industry', in *Proceedings of the 1984 International Conference on Occupational Ergonomics*, Toronto, Vol. 2, pp. 134–40.

Noro, K., and Imada, A. S., 1984, 'Participatory ergonomics', a workshop presented at the First International Symposium on Human Factors in Organizational Design and Management, Honolulu, Hawaii.

Noro, K., Kogi, K. and Imada, A. S., 1986, The current state of participatory ergonomics in the world, *Japanese Journal of Ergonomics*, **22**, 208–209 (in Japanese).

O'Brien, D. D., 1980, New products: harmonious designs come from well managed and coordinated teams, *Planned Innovation*, **3**, 111–14.

O'Brien, D .D., 1981. Designing systems for new users, *Design Studies*, **2**, 139–50.

Peabody, G. L., 1971, Power, Alinsky, and other thoughs, in Hornstein, H. A., Bunker, B. B., Gindes, M. and Lewicki, R. J. (Eds) *Social Intervention: A Behavioral Science Approach*, pp. 521–32, New York: Free Press.

Perrow, C., 1981, Disintegrating social sciences, *New York University Educational Quarterly*, **12**, 2–9.

Platten, P. E., 1984, The investigation of organizational commitment as a source of motivation in quality control circles, *Dissertation Abstracts International*, **45**, 708B.

Rafaeli, A., 1985, Quality circles and employee attitudes, *Personnel Psychology*, **38**, 603–15.

Seybolt, J. W. and Johnson, R. L., 1984, Monitoring the impact of quality circles: an example from Tenneco, *Transactions of the Sixth Annual IAQC Conference*, **6**, 152–6.

Seybolt, J. W. and Johnson, R. L., 1985, The effectiveness of quality circles at Tenneco: two years later, *Transactions of the Seventh Annual IAQC Conference*, **7**, 148–52.

Shores, D. L., 1985, An exploratory study into the relationship between quality circles and job satisfaction, *Dissertation Abstracts International*, **45**, 2014A.

Srinivison, C., 1983, Influence of quality circles on productivity, group behavior, and interpersonal behavior: an exploratory micro-organizational development perspective, *Dissertation Abstracts International*, **43**, 3055A.

Swezey, R. W., 1987, Design of job aids and procedure writing, in Salvendy, G. (Ed.) *Handbook of Human Factors*, pp. 1039–57, New York: Wiley.

Toffler, A., 1980, *The Third Wave*, New York: Bantam.

Tortorich, R., Thompson, P., Orfan, C., Layfield, D., Dreyfus, C. and Kelly, M., 1981, Measuring organizational impact of quality circles, *The Quality Circles Journal*, **4**, 24–34.

Waldron, I. and Jacobs, J. A., 1988, Effects of labor force participation on women's health: new evidence from a longitudinal study, *Journal of Occupational Medicine*, **30**, 1977–83.

Weick, K. E., 1984, Small wins. Redefining the scale of social problems, *American Psychologist*, **39**, 40–9.

Weick, K. E., 1987, Organizational culture as a source of high reliability, *California Management Review*, **24**, 112–27.

Wolfe, P. J., 1985, Quality circle intervention: structure, process, results, *Dissertation Abstracts International*, **45**, 2184A.

Yankelovich, D., 1979, Work values, and the new breed, in Kerr, C. and Rosow, J. M. (Eds) *Work in America: the Decade Ahead*. New York: Von Nostrand Reinhold.

Zahra, S. A., 1982, An exploratory empirical assessment of quality circles, *Dissertation Abstracts International*, **43**, 2030A.

Zander, A., 1979, Research on groups, *Annual Review of Psychology*, **30**, 417–52.

PART II
CASE STUDIES

Chapter 3

Participative Approaches to the Design of Physical Office Work Environments

R. G. Rawling

Introduction

Much has been written about the desirability of systems design because of the priority attached to the needs of people when deciding how work is to be structured. The argument is that by making appropriate and timely decisions about organizational aspects of job design, physical work environment factors and training/skill needs, the relationship between people and their work will be optimized. These goals cannot really be argued with; however, the application of ergonomics in this context often tends to focus too much on technical aspects, tending to overlook effective management of change.

Introduction of even simple anthropometric considerations into the selection of equipment can go astray if users are not convinced of the value of the refinements offered. With more sophisticated applications of ergonomics, perhaps where users cannot see any direct relevance to themselves, the challenge of successfully introducing changes is considerably greater. Those responsible for introducing better ergonomic practices into organizations may adopt one of two general approaches:

1. coercive change; or
2. participative change

Coercive change is characterized by use of authority leading to behaviour changes which have an inherent instability. It is uncertain how well the change will persist if the authority is removed.

Participative change is characterized by desired behaviour changes occurring as a result of group participation in deciding the direction and value of proposed changes. Inevitably, the time taken for participative change is slower, yet it is more likely that desired outcomes will persist.

The introduction of information technology into organizations is an area where the concepts of ergonomics, system design and participative change management are all highly relevant. Although, as noted by Eason (1982), there has been an apparent reluctance by industry fully to embrace computer technology, with an overly strong focus on simply acquainting people with the technology being an important limiting factor. It has certainly been our experience that technical knowledge is a necessary but insufficient condition for the successful introduction of information technology. Not only does it leave users pondering over the ability of the system to meet their needs, but this push in isolation from other considerations can create negative organizational ramifications, relating primarily to human resource issues.

It was with respect to employee health-and-safety impacts that the State Electricity Commission of Victoria (SECV) entered into a joint agreement with the major clerical employees association representing SECV employees to establish suitable physical working conditions when using screen-based equipment (SBE). Indeed, a process which started out with a focus on health-and-safety issues has now matured into a consideration of broader impacts relating to technological change. How the organization has managed these considerations using bipartite participative mechanisms is the focus of this chapter.

Information technology at the SECV

Computers are involved in almost every facet of the SECV with roles in system control, as an aid to decision-making or data retention. In administration, computers are used for personnel-records management, customer records, stores provisioning, and many other applications. In operations, computers assist in controlling power-generation plant and drive remote system telemetry devices. In design, computers are used to perform detailed engineering simulations and assist in drafting. In finance, computers assist in corporate planning and general econometric modelling.

The types of hardware used fit into three broad categories:

1. mainframe network devices (e.g. time-sharing users);
2. stand-alone systems (e.g. personal computers); and
3. plant control (e.g. control room devices);

There are predominantly three major categories of user:

1. information (text and systems data entry);
2. dialogue (involving both the SBE and hard copy); and
3. data inquiry (technical).

Support for these user groups, and the technology they use, is by and large provided by a central services group. Many users, with specialized applications, are self-supporting or have close liaison with relevant design groups. Aspects of office systems design, and general information systems development are vested in corporate support groups.

Employee relations approach

The SECV has a corporate employee relations objective which aims to foster a safe, satisfying and non-discriminatory work environment which enables all employees to contribute toward achieving business objectives. To facilitate attainment of this objective, broad strategies are being pursued which aim to:

1. encourage a participatory management approach, facilitating communication, and productivity gain;
2. provide effective accountability and responsibility for human-resource management;
3. take a responsive approach to social legislation such as health and safety and EEO;
4. provide effective resourcing of strategic business operations;
5. provide security of employment; and
6. provide comprehensive employee development and training programs.

In accordance with contemporary human-resource-management practice, the SECV clearly believes that both employee and corporate goals are interdependent. The introduction of appropriate employee participation initiatives can, therefore, lead to better cooperation, greater organizational effectiveness and improved work satisfaction for employees.

Health and safety

A specific example of the application of the employee relations philosophy outlined above is the SECV's agreement with unions on health and safety. These joint agreements now characterize the conduct of health-and-safety programs in many Australian organizations, and are underpinned by a legislative framework which establishes statutory provisions relating to workplace consultation on health-and-safety matters.

The SECV agreement has two general principles. First, existing standards, such as those produced by Government authorities, are regarded as minimum acceptable levels. Second, SECV management and employees and the relevant unions work together at all levels to develop, maintain and improve the provision of healthy and safe working conditions.

The agreement provides for the election of health-and-safety representatives and the establishment of both workplace Health and Safety Committees and an overall Health-and-Safety Policy Committee. Health-and-safety representatives are elected by the employees and it is a requirement that they are members of an appropriate trade union/association.

In each agreed workplace, health-and-safety matters are dealt with through a joint management–employee Health-and-Safety Committee. These committees have as their terms of reference the consideration of any matter relevant to employees' health and safety raised by management or employees.

The terms of this agreement relate just as much to workers in office environments as they do to traditional industrial tasks, indeed a large proportion of the workforce is engaged in 'office' work. It is particularly relevant that about the time that the Health-and-Safety Agreement was introduced, 'office' workers via their union were urging the SECV to implement standards relating to office working environments. The stage was therefore set for the introduction of a major new policy initiative to be managed in a bipartite fashion.

Development of a code of practice for the installation and operation of screen-based equipment

The SECV has a history of developing internal codes of practice to assist in the effective management of a variety of occupational health-and-safety issues (e.g. hearing conservation, and asbestos). Ordinarily, a process of line-management consultation facilitated by the corporate Health-and-Safety Department would have been undertaken to generate such Codes of Practice. However, coincident with growing SECV interest in the area of office working environments was a similar interest by the main SECV Clerical and Administrative Union (Municipal Officers' Association (MOA)), the members of which operate a large proportion of the SECV SBE. This interest was, to a certain extent, fostered by the Australian Council of Trades Unions/Victorian Trades Hall Council Occupational Health-and-Safety Unit (ACTU/VTHC), which has promulgated guidelines in this area. This background, combined with industrial negotiations concerning the design of jobs in a planned computer-assisted telephone-inquiry operation, precipitated consideration of a joint Code of Practice. This was a novel advance in participative approaches to occupational health and safety in Australia, and has since been adopted by other organizations.

The ACTU/VTHC guidelines and policy on SBE have been criticized for not prioritizing or defining the application of recommended equipment criteria. (Anon., 1983). It was suggested that the criteria were not practical as a guide for labour negotiations, as they were proposed to be applied on a constant basis for all classes of users, irrespective of circumstances. This criticism could also have been levelled at the MOA policy. It was clear that unqualified embracing of the MOA proposals could have created difficulties in line departments.

To enable meaningful negotiations to proceed, the corporate SECV Health-and-Safety Group prepared a draft Code of Practice based largely upon contemporary ergonomics guidelines (c. 1983). However, as is discussed later, subsequent renegotiation of these guidelines has proved necessary to enable more flexibility in line-management operations.

Issues arising during negotiation of the Code of Practice

Accommodation of joint concerns was relatively simple in relation to technical questions which were supported by scientific evidence, e.g. visual environment design and screen design. It was not as simple to obtain agreement where there was a lack of supporting scientific evidence. For example, it was the position of both the ACTU/VTHC and the union that there was an association between birth abnormalities and pregnant employees exposed to radiation whilst operating SBE, despite the lack of sound prospective epidemiological studies in support of this view. Whilst it has led to some criticism of the SECV, the decision to accede to the inclusion of clauses in the Code of Practice relating to such matters has spared the SECV a degree of industrial disputed experienced by some other Australian organizations.

A more important concern during the development of the Code of Practice was the achievement of a proper balance of content between technical concerns and systems design. Originally the Code of Practice was written with emphasis on technical concerns which almost certainly would have led to the perpetuation of a classic rather than the desired systems ergonomics approach. Emphasis would have been placed on the supply of ergonomically designed hardware and physical work environments. These matters, whilst important, have been shown by Sauter *et al.* (1983) to have little impact on reduction of visual and musculoskeletal stress. Questions of social support and job control, as related to job design, are also significant. The intent was to introduce the reader of the SECV Code of Practice to the concept of systems design. That the Code of Practice largely failed to establish an appreciation of this important issue is a function of the lack of understanding by both users and the organization of the relationship between the human and technical systems and the decision-making processes being used at the SECV at the time. This situation has now changed whereby technological change is introduced under the auspices of a public sector technological change agreement (see the section on participative technological change and ergonomics for further details).

The specific content of the Code of Practice is indicated in figure 3.1.

Figure 3.1. Structure of the initial Code of Practice.

1. *Introduction*
   Application of guidelines
   Sources of information
   Definitions
   Systems design
   Regulations and standards

2. *Work organization*
   Rest breaks and exposure time
   Bonus systems and monitoring of performance

3. *Education and training*

4. *Medical assessment*
   Visual factors
   Eye health
   Pregnancy
   Medical records

5. *Screen-based equipment characteristics*
   General
   Equipment appearance
   Construction of unit
   The screen
   Keyboard/screen assembly
   Slope of screen face
   Display
   Character, size and shape
   Screen and character colour
   Screen refresh rate and tube storage delay
   Contrast, focus and brightness adjustments
   Screen size
   Keyboard
   Keyboard layout
   Key and keyboard profile
   Key size and travel
   Colour and colour coding of keys
   Key stroke, travel, force, feedback and noise
   Error alarms and cursors
   Electrical safety
   Radiation standards

6. *Work-station and environmental assessment*
   Workplace layout and posture
   Terminal function types
   Workplace layout
      Viewing distance
      Seated posture
      Working heights
   Practical design of SBE workplaces
   Chairs
      Chair seat
      Chair back support
      Arm rests
      Chair base and castors
      Chair upholstery
      Chair adjustments
      Chair maintenance
   Work-station design
      Work-station top
      Work-station height
      Work-station design safety
      Work-station adjustability
   Footrests
   Document holders
   Environment
   Lighting
   Visual fields
   Ambient lighting levels
   Reflections
   Environment conditions
   Noise
   Decor and building finishes
   Safety
      Tables
      Appliance testing
      Passageways
      Fire alarms
      Static electricity in carpets

7. *Other*
   Tender specifications
   Glossary of terms
   Technical references

Implementation of the Code of Practice

Rawling and Shahbaz (1983) have noted that the development of standards, in the absence of effective implementation, may result in the aptly named 'in-basket syndrome', i.e. the initiative fails to be adopted. This is basically what happened with the SECV/MOA Code of Practice. Initially, the agreed Code of Practice was used to examine the aforementioned customer-inquiry network. Staff of the Corporate Health-and-Safety Group were involved in assessing proposed work sites for this network, and recommending actions for upgrading, based on application of professional judgement.

It was this *modus operandi* that turned out to be a major problem for the Health-and-Safety Group. Line departments clearly believed that responsibility for the Code of Practice lay with the Corporate Health-and-Safety Group. This group was also seen as arbiters of the Code and, more importantly, there was an expectation that the Corporate Health-and-Safety Group would implement the Code. By failing to establish responsibility and accountability for the Code within line departments, a true commitment to implementation was never really evident.

On the relatively rare occasions when involvement of the Occupational Health-and-Safety Group was sought, it was often amidst industrial dispute. Inevitably, in these situations a quick fix was desired and this centred on adjustment to physical factors. As a consequence, there was perpetuation of the belief that modifications to physical factors were sufficient.

For a period of 1 year, the Code of Practice was not adopted by line departments to the extent necessary to claim effective implementation. In early 1984, the union indicated a concern for this lack of implementation and initiated work bans on SBE work stations which did not comply with the Code of Practice. Initially, an increase in resources of the Corporate Health-and-Safety Group to assist in implementation was requested, although because of the critical need to establish ownership of the Code by line departments this push was resisted. The question was one of effective management of the implementation process, *not* one of importing more experts.

A subtle change in direction occurred between the initial release of the Code and imposition of bans (the 'in-basket' period). Initially, the Code was written as a set of guidelines as well as providing design criteria to be aimed at. Unfortunately, as a result of the failure to implement the Code effectively, a tendency emerged to view the Code as a set of *regulations*. In these circumstances, the Code was being applied on an essentially equal basis for all classes of users. Precisely the situation sought to be avoided in the first instance.

An instructive example of this relative inflexibility was the matter of 80-character vs. 132-character screen formats. A survey of some existing terminals indicated non-compliance with Code requirements as indicated in table 3.1. Two observations are relevant. Firstly, in the 80-character mode the terminals do not comply in all respects; however, the geometry approaches the ideal. Secondly, in the 132-character mode a reduction in all horizontal geometry is noted, with the most noticeable reductions being upper-character width and inter-character spacing. This situation was sufficient to result in an industrial dispute concerning the use of the 132-character mode because it did not completely comply with the Code of Practice requirements.

After extensive discussions, and given that *readability* was the prime aim of the original provisions in the Code of Practice, a decision was made to adjust raster spacing when in the

Table 3.1. Screen-based equipment code character geometry requirements and survey results.

Parameter*	SECV/MOA Code	Character mode 80	Character mode 132	Optimized recommendation
Upper-character height, H (mm)	3.1–4.2	4.30	3.60	3.1
Upper-character width (%H)	70–80	65	50	58
Stroke width (%H)	10–12	9.3	11	13
Inter-character space (%H)	20–50	14	7	8
Inter-word space (%H)	50–70	93	58	68
Inter-row space (%H)	100	84	90	105

* %H = percentage of upper character height H.

132-character mode to achieve the optimized geometry given in table 3.1. Whilst not ideal, it was considered that text familiarity would enable adequate readability given this adjustment. The issue demonstrated by this example is that hard scientific criteria (about which there are differing opinions in any case) cannot be dogmatically applied, particularly when industrial dispute is prevalent.

More effective implementation — change management

Clearly, the situation outlined above could not continue if business operations of the SECV were not to be severely restricted. To resolve the problems which had emerged, the matters listed in table 3.2 needed to be addressed. These forces had to be taken into account in the selection of a more effective implementation strategy. In addition, organizational responses needed to be monitored to ensure the maintenance of any implementation and compliance with the Code of Practice. The basic approach adopted centred around the concept of *project management*. However, in the context of change management, persons introducing ergonomic standards, such as those discussed in this chapter, should be aware of just how an overall change strategy affects outcomes.

Table 3.2. Factors considered in resolving the problems that arose in implementing the code of practice at the SECV.

Forces for change	Forces against change
Employee relations philosophy	Lack of implementation strategy
Health-and-safety agreement	Lack of resources
Community awareness	Lack of understanding
Union concerns	Role of corporate health-and-safety group

The existing organizational climate and concern about the resource requirements necessary to implement the Code of Practice made any use of *positive reinforcement* almost impossible. *Extinction*, whereby the lack of any organizational response extinguishes behaviour, would also not result in effective implementation of the Code of Practice because the prevailing view had already been that non-compliance with the Code resulted in no discipline. Usual practice was to get the technology, and leave someone else (e.g. the Health-and-Safety Group) to worry about negative organizational ramifications. The only remaining approaches open at this time were, unfortunately, *negative reinforcement* and *punishment*. These approaches unfortunately included threats of bans and non-operation of equipment and to commence the project in this negative way required considerable finesse. It is a salutory notice to all occupational health-and-safety practitioners that a proactive approach to workplace issues is highly desirable.

Project-management approach to implementation

The experience of using project management in this implementation program clearly indicates that this approach is one which many more ergonomics practitioners should adopt in facilitating the introduction of programs. This approach is particularly suited to participative programs because it has the potential to provide for effective input from union and employee representatives.

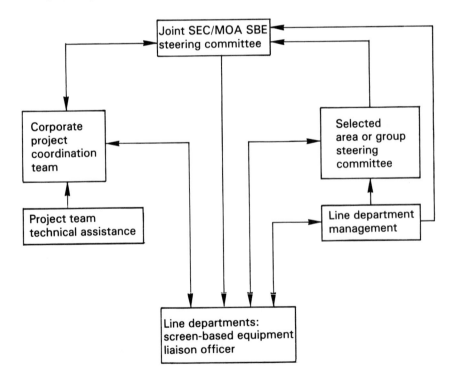

Figure 3.2. Project-management framework for participative implementation of the Code of Practice.

In the case study described in this chapter, the structure presented in figure 3.2 was jointly adopted by the SECV and the unions. The key elements of this structure included:

1. a joint SECV/Union Corporate Steering Committee;
2. a Project Coordination Team (including the Project Manager); and
3. Project Team technical assistance.

Joint SECV/Union Steering Committee

The joint SECV/Union Corporate Steering Committee was formed to oversee and coordinate implementation of the Code of Practice. This committee comprised representatives from both the union and the SECV line management. Representatives from line management were taken from those departments with the greatest number of SBE installations. This committee met monthly during the renewed implementation effort and developed a wide variety of implementation guidelines. Many of these jointly agreed guidelines on *how* to implement the Code have persisted and are reflected in a renegotiated Code (see section on effectiveness of implementation).

Project Coordination Team (including Project Manager)

As the implementation of the Code of Practice was both a technical and an industrial-relations matter, the Corporate Human Resources Directorate decided on a Project Coordination Team approach. The group was led by a Project Manager, a technical expert on the Code who was assisted by industrial relations and administrative support.

Project Team technical assistance

A technical-support team of six employees seconded on a full-time basis was formed to deal with the backlog of assessments of SBE work stations. This team undertook an initial period of training and then carried out assessments of work stations in individual departments. The assessment involved reviewing the visual, acoustic, thermal and general environment as well as undertaking a reading of SBE radiation emissions. The operation of this team was novel because none of the members knew anything about ergonomics in general or the applications of ergonomics to SBE. The team used an annotated check-list to assist with assessments (see below).

Line department networks

Each line department nominated a representative to act as a SBE liaison officer. The representatives were responsible for facilitating and coordinating the implementation of the Code within their respective areas. These representatives participated in two 1-day seminars designed to provide a background to the Code and discuss aspects of the implementation process.

Apart from people resources, three other important techniques were adopted to facilitate implementation:

1. annotated workplace assessment check-lists;
2. performance measurement; and
3. education and training.

Figure 3.3.   Examples of the general, visual and acoustic environment check-lists used in the performance measurement and control.

Department _____   Division _____

Section _____

Physical Location _____

Serial No_____   Terminal Type_____

Type of Use–   Word Processor   ____   Micro/Personal   ____   TSO   ____
               Data Processor   ____   CAD   ____   CES   ____
               Stores   ____   Programmer   ____   Computer Operator   ____
               Other   ____

Relevant SBELO   [_____]   ext [_____]

Responsible Person   [_____]   ext [_____]

Attachments

	Yes	No	Comment			Yes	No	Comment
visual environment	___	___	_____	acoustic environment		___	___	_____
radiation	___	___	_____	thermal environment		___	___	_____
general environment	___	___	_____	work-station photograph		___	___	_____

PLACE REVIEWED STICKER ON SBE

Suggested Action
VISUAL/GENERAL _____
_____
_____
_____
_____
_____
_____
_____
_____

THERMAL _____   RADIATION _____
_____   _____
_____   _____

ACOUSTIC _____
_____
_____

WORK STATION (detail optional) _____

ACTION   Forward the above, together with attachments to SBE project
         coordination team.   Signed _____
                                                               Date _____

Occupational Health Division   Checked   OK ☐   NOT OK ☐
Signed _____   Date _____
Comment _____
_____

Action for SBECO
   Lighting Services ☐   Noisecon ☐   Furniture ☐   Suggested Option No ☐

Operation Authority   (optional)
                      Lighting ☐   Furniture ☐   Noisecon ☐
(If yes to above action)
SBELO ☐   User Representative ☐

SBE ASSESSMENT MODULE REVISION ACOUSTIC WORKING ENVIRONMENT

PARAMETER	RECOMMENDED	MEASURED	ACCEPTABLE? Yes	No	COMMENTS
1 L10 (dBA)	< 63				Measured at operator's position. Paste readout to back of this page.

Noise sources in the area:					Printer has acoustic enclosure? Yes ☐ No ☐
1 Printer (1)	≤ 53 dBA (s)				Measure all levels at a distance of 1 m and a height of 1.2 m and when the machine is in operating mode. If levels vary state ranges
2 " (2)	" " "				
3 Photostat Machine	" " "				
4 Typewriter (1)	" " "				
" (2)	" " "				
5 Other (1)	" " "				_____
" (2)	" " "				_____ _____ _____

Octave Band Levels:	dB (slow)				Measure octave bands at position of operator and only if L10 > 63 dBA. If background levels vary, state ranges of levels.
63 Hz	74				
125 Hz	68				
250 Hz	63				_____
500 Hz	58				_____
1 KHz	56				_____
2 KHz	54				_____
4 KHz	53				_____
8 KHz	52				_____

Reverberation

Room appears

Very Reverberant			Good		Dead	
1	2	3	4	5	6	7

- Does the room appear to have echoes? Yes ☐ No ☐
- Does the sound seem to travel easily? Yes ☐ No ☐

Headsets

- In use? Yes ☐ No ☐
- In use for about _____ hrs/day on average.
- Type: _____
- Gain control? Yes ☐ No ☐
- Headset used for _____

Masking

- Do the operators find it difficult to communicate   Yes ☐ No ☐
- Is the difficulty experienced, face-to-face;   Yes ☐ No ☐
-    over phone;   Yes ☐ No ☐
-    other, specify _____   Yes ☐ No ☐
  _____   Yes ☐ No ☐
- What appears to be the cause of the difficulty?
  _____

Metrologger No _____ No _____ No _____

Reader No _____ SLM _____

Equipment calibrated before and after measurements?   Yes ☐ No ☐

Time(s) of day measurement carried out _____

Activity in area was: Average ☐   Above Average ☐   Below Average ☐

VISUAL WORKING ENVIRONMENT

PARAMETER	RECOMMENDED	MEASUREMENT		ACCEPTABLE		RECOMMENDATIONS
				Yes	No	
NOTE: All luminance measurements should be performed under horizontal task illuminance of 350–500 Lux. If not, note total (A + D) illuminance here _____ Lux.						
• Service Illuminance* (horizontal and at centre of keyboard)	no hard copy 200–350 lux hard copy plus	Lights on	1x			1
		off	1x			
		*Difference	1x			
• Luminance Balance (near field)		screen	cd/m2			2
1 Measure object luminance from eye position, and measure as found.		character	cd/m^2			
		bezel (bottom centre)	cd/m^2			
NOTE: Use close-up lens for character luminance (apply optical correction factor)		keys (brightness	cd/m^2			
		key surround (top centre)	cd/m^2			
		source document (note position)	cd/m^2			
		desk top	cd/m^2			
• Illuminance (near field)		bezel	1x			3
Measure object illuminance in plane of surface		key surround (top centre)	1x			
		keys	1x			
		desk top	1x			
		source document	1x			
• Luminance Balance (peripheral field)		anterior plane	cd/m2			4
1 Measure object luminance from eye position		posterior plane	cd/m^2			
		lateral plane	cd/m^2			
		floor	cd/m^2			
		ceiling	cd/m^2			
		window	cd/m^2			
• Illuminance (peripheral field)		anterior plane	1x			5
1 Measure object illuminance as above		posterior plane	1x			
• May be impossible		lateral plane	1x			
		floor	1x			
		ceiling*	1x			

PARAMETER	RECOMMENDED	MEASUREMENT		ACCEPTABLE		RECOMMENDATIONS
				Yes	No	
• Luminance Balance (near field)	8:1 to 10:1	character – screen	:1			
2 Calculate contrast ratios by dividing lower *into* higher luminance	1:3	screen – bezel	1:			
	1:3 to 1:5	screen – keys	1:			
	1:3 to 1:5	screen – key surround	1:			
	1:3 to 1:5					
	1:5	screen – desk	1:			
	1:5	screen – document	1:			
• Reflectances (near field) Calculate reflectances	30–40%	bezel	%			
$R = \dfrac{100n \text{ Luminance}}{\text{Illuminance}}$	25–30%	keys	%			
	20–30%	Key surround	%			
	20–40%	desk top	%			
	60–70%	source document	%			
• Luminance Balance (peripheral field)	1:10	screen – AP	1:			
Calculate contrast ratios as above	1:10	screen – PP	1:			
	1:10	screen – LP	1:			
NOTE: Anterior and posterior with respect to operator	1:10	screen – floor	1:			
	1:10	screen – ceiling	1:			
	1:20 to 1:40	screen – window	1:			
• Reflectances (peripheral field)	40–50%	anterior plane	%			
2 Calculate reflectance as above	30–45%	posterior plane	%			
	60–80%	lateral plane	%			
	20–40%	floor	%			
	70–80%	ceiling	%			

VISUAL WORKING ENVIRONMENT (cont)

PARAMETER	RECOMMENDED	YES/NO	COMMENT	RECOMMENDATION
• Artificial Lighting Installation				
• diffusers	must be fitted and be of sharp cut off type			
• position with respect to SBE	line of sight must be parallel to long axis of light			
• adjustment	if adjustment by dimmer is provided, lights should not flicker <200 lux			
• visual rest opportunity	should be plane surface in immediate line of sight			
• glare-ambient	no sources of direct light in field of vision			
• keyboard and surrounds	glare patches on keys			
• documents	no expanses of white paper			
• Daylight				
• SBE position	line of sight parallel to windows			
• curtains	light scrim curtains advisable			
• sunlight	direct penetration to be filtered (e.g. awnings, solar film)			
Work Area				
Viewing Distance (from approximate eye point) Screen Keyboard	500–700 mm 450+ mm			
Hardcopy	support on document holder near to vertical		HOLDER – Present Not Present Appropriate Inappropriate	

Other Comments _____
_____
_____
_____

Workplace assessment check-lists

The workplace assessment check-lists lifted technical requirements out of the Code of Practice and placed them in a practical, action-orientated report format. The purpose of doing this was to assist line departments in applying the Code to their applications. In addition, the check-lists provided a basis for nominated union representatives to concur that a particular workplace had in fact been upgraded to the standards agreed to by the Steering Committee. Examples of the general, visual and acoustic environment check-lists are given in figure 3.3.

Performance measurement and control

Performance measurement and control techniques were adopted at two levels to assist the Steering Committee to manage the implementation process. To monitor progress toward upgrading approximately 900 SBE work stations, a measurable curve (as shown in figure 3.4)

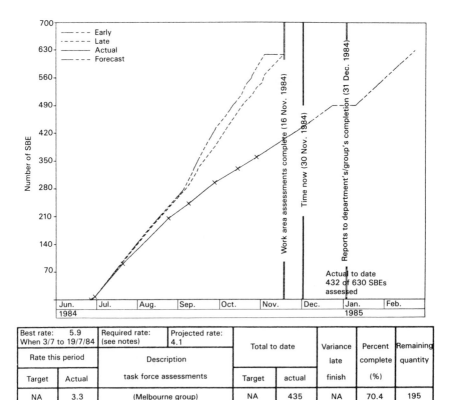

Best rate: 5.9 When 3/7 to 19/7/84	Required rate: (see notes)	Projected rate: 4.1	Total to date		Variance late	Percent complete	Remaining quantity
Rate this period		Description					
Target	Actual	task force assessments	Target	actual	finish	(%)	
NA	3.3	(Melbourne group)	NA	435	NA	70.4	195

Basis: Envelop of early start, late start is based on the Melbourne task force estimated and actual performance at 22 July 1984.

Quantity assessment derived from actual SBE assessment counts.

Notes: Forecast indicates that all assessments will not be completed until 23 February 1985.

Rates are given as number of SBEs per working day.

Figure 3.4.   Measurable curve used to monitor progress towards upgrading screen-based-equipment work stations.

was used to monitor positive and negative progress. Secondly, to manage the work-station upgrading process a system of stickers was used to indicate operational status. If the requisite upgrading work had been carried out to the satisfaction of both line management and employee representatives a clearance sticker was fixed to the computer terminal. In addition, a suitable entry was made in a register of all work stations which had to be maintained within each department. The stages in this clearance process are shown in figure 3.5.

Figure 3.5. Participative process adopted to monitor upgrading of individual screen-based-equipment work stations.

1. Departmental SBE Liaison Officers arrange for:
   (i) delivery of furniture and equipment,
   (ii) adjustment of lighting and noise levels to meet the Code of Practice,
   (iii) placement of furniture in a suitable location,
   (iv) the Technical Assessment Team to complete an assessment of the work station, and
   (v) all work necessary to bring the work station into compliance with the Code (as per recommendations of the Technical Assessment Team)
2. SBE Liaison Officer arranges for joint approval of the newly set up work station by contacting a union official
3. SBE Liaison Officer and the union official inspect the work station and check that all recommendations on the assessment have been carried out
4. Clearance sticker is attached to the work station to indicate that it meets the standards in the Code of Practice and is available for use

Education and training

Education and training also represented an important facet of the implementation program. The SECV and the union jointly developed a promotional video which used comedy and a story about convincing an elderly employee to upgrade his SBE work station. His work station was the last to be upgraded, and the video highlighted the joint approach between line management and a union representative in introducing the Code of Practice. This video did not concentrate on telling people how to adjust lighting, sit correctly and so on; it was aimed at inducing behaviour changes. The video was supported by suitable pamphlets and specific advice on posture, etc., where this was desired by employees.

Effectiveness of implementation

There are two levels on which to consider the effectiveness of implementation. The first relates to *ongoing application of the Code* and the second relates to the *framework for policy development*. Both these issues clearly relate to the issue of participation because both must be structured to accommodate this principle.

Ongoing application of the Code of Practice

The ongoing application of the Code has necessitated some rethinking of some of the processes established by the Steering Committee discussed earlier. The key changes which have been made include the following.

1. Establishing *five* new categories of SBE use (intensive or high time, mixed functions, intermittent, control rooms and technical) which more closely reflect the different problems relating to workplace layout, organization, environment and SBE configuration.
2. Modifying the Code in *all* respects to indicate *how* it shall apply to each of the above categories.
3. Firming up provisions relating to the broader issue of technological change (see section on participative technological change and ergonomics).
4. Confining the application of work pauses to intensive or high time users, unless in other situations session lengths exceed 50 min. A related initiative is the restructuring of screen-based and non-screen-based keyboard tasks to provide for multi-skilling.

Table 3.3.   Details of the weighted SBE terminal assessment check-list.

SBE Parameter		Weighting (%)
Display screen overall	30	
Phosphor type		23
Image stability		23
Screen reflections		23
Other		31
Character display	22	
Upper-character height		31
Inter-character discrimination		31
Other		38
Adjustability	17	
Character brightness		42
Screen angle		42
Other		16
Luminance balance	12	
Keyboard	18	
Home row thickness		38
Other		62
Other		1
Total		100

Category of use			
Intensive or high time	Mixed function	Intermittent	Technical
Better than 80% overall. Better than 85% on keyboard design	Better than 80% overall. Better then 80% on visual factors	Acceptable* on all aspects. Better than 80% on keyboard design	Acceptable* on all aspects

* Acceptable means a 50% score.

5. Inclusion of a range of procedures aimed at facilitating application of the Code requirements in situations such as:
     (i) new purchases of SBE,
    (ii) testing and evaluation of proposed SBE,
   (iii) new SBE work-station installations,
    (iv) relocation of SBE work stations,
     (v) clearance for operation,
    (vi) disabling of SBE (i.e. for non-compliance),
   (vii) audits of SBE workplaces,
  (viii) replacement of SBE,
    (ix) retirement of old SBE, and
     (x) assessment of areas containing multiple SBE.
6. Modification to a variety of technical criteria in response to practical experience and further ergonomic research at the time (e.g. adoption of a 1:5 screen: near-field-object contrast ratio, rather than the original 1:3 ratio).
7. Adoption of a weighted SBE terminal assessment check-list which better selects terminals in accordance with Code requirements and projected use. Details of this weighted assessment check-list are given in table 3.3.

Framework for policy development

The framework for policy development has also been critically reviewed to ensure effective input from line departments concerning Steering Committee decisions. This aspect of policy development is very important because participative-management approaches must operate in parallel with line-management structures. Traditionally, responsibility and accountability is vested with the line structure, whereas the participative structure has more of a monitoring/review role. Unless participative structures are held similarly accountable for the decisions taken, mechanisms such as those indicated in figure 3.6 may be necessary to maintain organizational effectiveness. The type of process depicted in figure 3.6 to a certain extent acts to manage the shifts in organizational power which can occur with participative approaches. Both in relation to the SBE Code of Practice and broader workplace issues, the SECV is not practising joint decision-making; however, it is firmly committed to effective and thorough employee participation and involvement.

Participative technological change and ergonomics

This case study relates to the quite specific impacts of a particular technology. The SECV also considers major technological change in a broader context and requires that employee impact statements be prepared taking into consideration:

1. the policies and directives of the Government;
2. the efficiency and effectivness of the organization, especially with respect to the envisaged costs and operations of the proposed change;

3. the effect on the staff, particularly in relation to
    (i) the number and type of staff required,
    (ii) training and retraining,
    (iii) redeployment, redundancy possibilities,
    (iv) promotion opportunities,
    (v) job satisfaction,
    (vi) skill levels and qualifications required, and
    (vii) occupational safety and health aspects and measures;
4. the service to clients and community;
5. the changes in the organizational structure and work arrangements;
6. the privacy of the system and civil liberties;
7. the effect on other organizations; and
8. other factors relevant to a particular technological change.

Employee impact statements which address the issues listed above are then considered as part of a broader process which adopts a *systems ergonomics* perspective. The process includes the following stages.

1. *Concept stage*: ideas concerning technological change first occur, are brought to the attention of senior management and are discussed.
2. *Contemplative stage*: resources are firmly committed to study the proposal.
3. *Development stage*: employees, consultants or any other persons are commissioned to carry out a detailed investigation or feasibility study of the proposed change and make recommendations to management.
4. *Approval stage*: management decides to proceed with the proposal and tender specifications are defined and issued.
5. *Implementation stage*: the technical change is introduced as either a fully operational, localized pilot or trial change.
6. *Post-implementation stage*: the operation and impact of the technical change is monitored and reviewed if necessary.

These developments, combined with the formal adoption of a *human resource planning* process is now moving the SECV away from a technology-led approach to decision-making. Human resource planning is a process whereby line departments are required to indicate on an annual basis how they plan to address a range of quantitative and qualitative human resource management issues. The success of these initiatives will rely on the extent to which it can be demonstrated that good human and technical resources planning is critical to overall success of the organization. This, in turn, requires the backing of senior management and the proper integration of planning processes (e.g. strategic, financial and resourcing).

With the project described in this case study, together with other human resource management initiatives, existing decision processes will be progressively modified to address relevant issues proactively. To do this requires that attention be given not only to input into the design process, but also to the way in which these inputs are provided. The creation of consultative forums, such as a Steering Committee, is one option; however, this is often accompanied by a feeling of loss of control amongst line managers. Invariably, the organizational support provided by a network of responsible persons is also necessary to deal with issues.

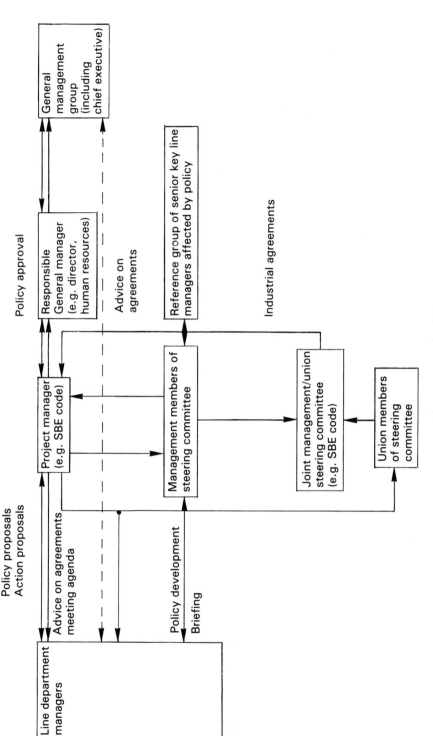

Figure 3.6. *Framework for participative policy development.*

It is likely that true systems design will only come about when designer-level technical decision-making is sufficiently influenced by human resource issues. At this stage it will be appreciated that simple environmental-modification solutions do not adequately address human resource concerns. The desire will then be for an approach to work and work-systems design which will automatically manage and develop human resources. Achieving the goal of successful systems design also clearly requires the involvement of employees and unions in the process.

Conclusion

It is clear that the development of ergonomics initiatives such as those described in this case study without proper attention to the implementation phase is fraught with problems. The SECV is of the view that it is possible and desirable to address both these issues in a proactive and participative manner. To do this requires the creation of relevant organizational support networks and decision-making structures. Correspondingly, the practicality of a variety of traditional ergonomic recommendations may need to be critically reviewed. It is often impossible to achieve fully in practice what is ideal in research. In this context, ergonomics practitioners should pay increasing attention to both the inputs they provide to the design process and the way in which these inputs are provided. In the long run, proper attention to human resources planning processes and the creation of true systems decision-making incorporating effective employee participation will be a necessary precondition for effective organizations.

References

Anon., 1983, Standards and legislation — ACTU health and safety policy on SBE, *Ergonomics Newsletter*, **2**, 2–8.

Eason, K. C., 1982, The process of introducing information technology, *Behaviour and Information Technology*, **1**, 197–213.

Rawling, R. and Shahbaz, C., 1983, 'Ergonomics and organisational effectiveness', in Proceedings of the 20th Annual Conference of the Ergonomics Society of Australia and New Zealand, Adelaide, pp. 293–308.

Sauter, S., Gottlieb, M. S., Jones, K. C., Dodson, V. N. and Rohrer, K. M., 1983, Job and health implications of VDT use — initial results of the Wisconsin – NIOSH study, *Communications of the ACM*, **26**, 284–94.

Chapter 4

Participatory Training for Low-Cost Improvements in Small Enterprises in Developing Countries

K. Kogi

Introduction

The important role of small and medium-sized enterprises in economic and social development is widely recognized. However, the conditions of work and the occupational hazards in the small-scale sector are causing increasing concern. It is difficult for the inspectorate and technical institutions to deal with all the problems of these widely dispersed enterprises. Moreover, it is difficult for the enterprises themselves to take action, because they have pressing financial, marketing, infrastructure and operational problems.

Despite these constraints, there is a growing awareness that the improvement of workplace conditions from ergonomic and other relevant points of view is possible through the active participation of managers and workers (Imada, 1985; Kogi, 1985b; Wisner, 1985). With direct support of training at various levels, many improvements can be brought about at workplaces in developing countries (WHO, 1981; Kogi, 1987a; Kogi and Sen, 1987).

Recent experience within the framework of the International Programme for the Improvement of Working Conditions and Environment (PIACT), launched by the International Labour Office (ILO) in 1976, shows that small and medium-sized enterprises can realize concrete improvements when training is participative and makes full use of voluntary action. This experience has also shown that direct support should be provided to enterprises, particularly for going through the process of learning-by-doing (ILO, 1985, 1986, 1988).

The types of support for facilitating participatory training are discussed on the basis of recent experience. If appropriate and direct support is given to small enterprises, many improvements can be made by means of self-help action using existing resources.

New training approach

To overcome the difficulties inherent in small enterprises, particularly those in developing countries, it is essential to develop training approaches that depend on voluntary cooperation rather than enforcement and thus help each enterprise carry out immediate improvements. Small enterprises need to be backed by well-organized support that provides technical and management training and advice (Bromley, 1985).

The critical questions in making workplace improvements concern: the identification of priority problems in the local context; and the implementation of effective solutions at low cost (Dy, 1984; Kogi, 1985a). Therefore, support should be organized so as to enable people to find priority problems and effective solutions using locally available materials and skills (Sen, 1984; Kogi and Sen, 1987). This support should provide for:

1. practical advice on how to identify priority problems and how to find solutions (about 'how-to' and not 'you must'); and
2. concrete guidance, particularly by learning-by-doing, about implementing immediate improvements (for 'self-help').

An example is given by a new training programme which has been developed within the ILO's PIACT activities. The target groups are entrepreneurs and workers of small and medium-sized enterprises (Louzine, 1982). The programme represents a systematic approach to the simultaneous improvement of working conditions and productivity in these enterprises (ILO, 1985, 1986, 1988; Kogi, 1987b). The approach is designed to encourage and assist small enterprises to take low-cost, voluntary measures which improve working conditions while at the same time increasing productivity.

This training programme is voluntary and participatory, and is carefully adapted to local circumstances. Pilot courses have been successfully conducted in India, Indonesia, the Philippines, Thailand and Argentina. They have demonstrated that small enterprises which have been exposed to this approach carry out substantial improvements which have clear benefits for both employers and workers.

In organizing training using this approach, several options are possible. An intensive training course is generally carried out over 4 weeks, while a short training course of 1 or 2 days can also be organized. An intensive course consists of: a preparatory phase for collecting information and slides from participating enterprises; a checklist exercise; technical

Table 4.1.   Three-month follow-up results of improvements suggested at the end of a small enterprise training course in Madras, India, in 1985. Results of 27 enterprises.

Types of action	Number of improvements		
	Suggested	Done or in progress	To be done soon
Good housekeeping	19	14	3
Better materials handling	8	4	3
Changing work height or machine designs	5	4	1
Improved lighting	11	9	2
Screening hazard sources or providing exhausts	11	11	
Providing guards or protective equipment	6	4	2
Reducing fire hazards	2	2	
Sanitary facilities or lunchrooms	8	8	
Total	70	56	11

workshop sessions dealing with several topic areas to discuss examples from these enterprises; mutual enterprise visits by participant groups and preparation of action plans; and improvements in enterprises and preparation of group reports to a final workshop.

From such courses, a series of improvements has resulted. On average, three to five improvements per enterprise were implemented over a 3-month follow-up period. Table 4.1 gives the results of a course in India which included a range of simple improvements. About 80 per cent of the measures proposed were implemented within the 3-month follow-up period (Kogi, 1987b). It is important to note that these improvements were designed voluntarily and realized without outside help.

Training principles and technical content

Based on these pilot activities, it has been established that such training can lead to many, concrete improvements when the six principles listed in table 4.2 are followed (ILO, 1988).

It is particularly important to build on local practice and to focus on achievements. Thus, the collection of positive examples obtained directly from the participating enterprises and their use in training sessions are essential. These positive examples, collected in the form of slides, show clearly what can be done and how improvements can be carried out. Since the examples are taken from the workplaces in similar conditions, they can be used as the actual basis for improvement action. Parallel to discussing these examples, the participants are guided to check, using a check-list consisting of about 50 simple, low-cost measures, which can be applied to their own workplaces.

Table 4.2.   *Six basic principles followed in an ILO training programme for higher productivity and a better place to work in small and medium-sized enterprises.*

1.   Build on local practice
2.   Focus on achievements
3.   Link working conditions and other management goals
4.   Use learning-by-doing
5.   Encourage exchange of experience
6.   Promote workers' involvement

In the subsequent small-group work, the participants mutually discuss these potential measures. The participants then decide to carry out some of them within the course period, usually within a few days or within a week. Group presentations are prepared based on these improvements. In the final workshop, the participants present those improvements which have already been carried out and those which can be completed within a 3-month period. In this way, the learning-by-doing process actually facilitates the implementation of possible changes.

The technical content of each training course depends on the specific problems and opportunities found in participant enterprises. However, it seems useful to concentrate on those topics which are common to most enterprises and which have a direct impact on productivity. This is extremely important in view of the voluntary nature of the course. Productivity-related topics can attract the attention of managers as well as workers and are, in effect, crucial for improving the work methods and the working environment.

Thus the technical content of the course is oriented toward solutions rather than to the problems themselves. Slides from the participant enterprises are selected to give good examples of practical, low-cost solutions deserving wide application. According to the training experience within PIACT, the technical content can best be organized around the eight technical themes listed in table 4.3. These themes have been chosen because of their practical importance in small and medium-sized enterprises and because of their relationship to both working conditions and productivity.

Table 4.3. Eight technical themes, each selected for the ILO's small-enterprise training courses because of its practical importance and its relationship to both working conditions and productivity.

1. Materials storage and handling
2. Work-station design
3. Productive machine safety
4. Control of hazardous substances
5. Lighting
6. Welfare facilities and services
7. Work premises
8. Work organization

According to the ILO's introduction to this training programme (ILO, 1988), these technical themes have the following features and benefits.

1. *Materials storage and handling*, such as use of storage racks and stands, use of push-carts, leverage or mechanical aids and better materials-moving and -lifting methods. Improved materials storage and handling mean recovery of misused space, less production time spent searching for tools and materials, lower capital costs due to less work-in-progress, simplified inventory control, fewer unnecessary operations and a better overall factory appearance.
2. *Work-station design*, such as placement of components within easy reach, better feeding methods, use of holding devices, use of platforms or chairs and clear displays on instructions. The better postures and motions thus achieved mean less fatigue and higher productivity and quality.
3. *Productive machine safety* including fixing guards and handrails, interlocks, improved and safer work methods and ready access to emergency controls and devices. It is often possible to increase productivity while at the same time eliminating the hazard.
4. *Control of hazardous substances* including better handling and storage of hazardous substances, local exhausts and enclosure. Simple and inexpensive measures can control most exposures to these substances whilst at the same time ensuring more efficient operations, preventing spoiling of materials or products, and reducing extra inspections or absenteeism.
5. *Lighting* such as the use of skylights and better lighting arrangements. Better lighting and related visual improvements very often increase productivity and reduce rejects.
6. *Welfare facilities* such as provision of drinking water, eating places and good sanitary facilities and arrangements for meals, first aid and recreational facilities. Good welfare facilities improve workers' health, morale, job satisfaction and attendance.

7. *Work premises* such as better planned layout, increased natural ventilation, isolation of sources of heat, noise, dust and fumes or clearing and marking of passageways. These lead to better flow of materials, improved ventilation, protection from heat and pollution and better use of available space.
8. *Work organization* such as more frequent changes in tasks, use of buffer stocks, arranging less stressful operations or relay-out of operations. The results are smoother and more efficient work flow, reduced down-time, reduced need for supervision and better use of worker skills.

Improvements through self-help

The improvements accomplished through our training courses have some common characteristics (ILO, 1985, 1986, 1988). Firstly, almost all the improvements made are simple, low-cost solutions. Secondly, the improvements are designed voluntarily as a result of learning of similar examples. Thirdly, strong linkages can be identified between these improvements and benefits. Usually the improvements result in visible benefits in terms of productivity and working conditions. In other words, these improvements have been favoured because they can be beneficial for both employers and workers. Nevertheless, as a rule, the changes made represent important improvements in safety, health and welfare. Finally, an enterprise usually takes action in multiple areas. The participants see benefits in multiple areas, apparently learning from the examples used in the course.

These characteristics point to the importance of the self-help nature of the improvements. If the managers and workers do not see any likelihood of a productivity gain and do not learn to use their own ideas and skills, they will quickly lose interest. Therefore, it seems essential to use good training tools that can stimulate and sustain the self-help process. The training tools used in the main elements of the training course are indicated in table 4.4. Three main types of training tool prove useful.

One important tool is the check-list. The check-list we have used consists of simple, low-cost measures which cover all the technical areas shown in table 4.3. The participants use the

Table 4.4. *Training tools used effectively in a training course based on the new approach.*

Training element	Training tools
1. Initial enterprise visits	Interviews and taking photographs
2. Joint check-list exercise	Check-list
3. Technical workshop sessions	Slides showing good, local examples
4. Case studies	Slides and illustrations showing the situation before and after improvement
5. Action in small groups	Group discussion based on the results of the application of the check-list and action planning forms
6. Group presentations	Slides and illustrations showing the situation before and after improvements

check-list to choose those items which they think are applicable to the workplace in question. In this sense, the check-list is an action check-list and different from normal types of check-list which ask the users whether the condition is appropriate or not. The type of check-list we use guides the participants to select the measures to be taken in a very direct manner. Such a check-list seems very practical and suited to the action process. Examples of check-list items used in our training programme for small and medium-sized enterprises are given in table 4.5.

The second important tool is the use of slides obtained from the participating enterprises. Since the slides are of the participants' own enterprises, they represent local conditions.

Table 4.5. *Examples of check-list items from a check-list used in a training course on making enterprises more efficient and better places to work for small and medium-sized enterprises (ILO, 1988).*

	Do you propose action?			Remarks
	No	Yes	Priority	
1. Clear everything out of the work area which is not in frequent use	....	....	....	....
2. Provide convenient storage racks for tools, raw materials, parts and products	....	....	....	....
3. Use specially designed pallets to hold and move raw materials, semi-finished goods and products	....	....	....	....
4. Put stores, racks, workbenches, etc., on wheels for easy handling	....	....	....	....
5. Use carts, movable racks, cranes, conveyors or other mechanical aids when moving heavy loads	....	....	....	....
6. Put switches, tools, controls and materials within easy reach of workers	....	....	....	....
7. Use lifts, levers, or other mechanical measures to reduce the effort required by the worker	....	....	....	....
8. Provide a stable surface at each work station	....	....	....	....
9. Use jigs, clamps, vices or other fixtures to hold items while work is done	....	....	....	....
10. Adjust the height of equipment, controls or work surfaces to avoid bending postures or high hand positions	....	....	....	....
11. Change working methods so that the workers can alternate standing and sitting while at work	....	....	....	....
12. Provide chairs or benches of correct height with a sturdy back rest	....	....	....	....

Improvements illustrated by these slides are convincing as they have already been realized under the same local situation. The value of local slides cannot be over-emphasized. These slides should be used not for worst-case demonstrations but for the purpose of learning about successful cases. Good examples are of prime importance. The learning process is greatly facilitated by using slides showing the workplace scenes before and after improvement. (So far slides have been effective. In some cases, video-clips have also been useful. For practical reasons, however, slides are usually favoured.)

The third important tool is obviously the sharing of views and experience by means of small-group work, based on the results of the application of the check-list and the suggestions written down on action planning forms. These forms are used merely to note down for discussion the main action to be taken within the participant enterprises. What is noted down is usually similar to the low-cost measures listed in the check-list. As the participants are already guided to select simple, low-cost measures, they can easily identify the priority, workable improvements. In fact, they reach some solutions which are really innovative and interesting. Typical examples include an improved, safer feeding device using a gravity chute for feeding a grinding machine, a work-stand with sockets for hand-held electric tools, or a work station enabling workers to do the same work in either a standing or sitting position.

These tools can also be effectively used in a short course of 1–2 days. A short course may consist of a check-list exercise, technical sessions to discuss good local examples shown by slides, group work for suggesting improvements at some of the participant enterprises and final group presentations.

In short, it is extremely important to use local examples and to enable participants to find practical improvements by means of self-help. Audio-visual aids and check-lists are merely a means of facilitating this self-help process. Action manuals which illustrate low-cost solutions in main topic areas are also very helpful, if their contents correspond to the areas covered by the audio-visual aids and check-lists. Small-group work, which is the most important part of the training, should be carried out by the participants themselves, through helping each other and through making specific improvements in their own enterprises, taking up the suggestions made by themselves. Thus, the self-help approach can lead to a range of low-cost improvements.

Conclusions

Participatory training can lead to real improvements in small and medium-sized enterprises in developing countries. The training can enable managers and workers to identify priority problems and find solutions by providing practical advice and concrete guidance about implementation.

For effective training, it is essential to build on local practice and to focus on achievements. Simple, low-cost examples should be presented in a local context, preferably from the participants' own enterprises. The technical content should concentrate on themes which have practical importance and are related to both working conditions and productivity, e.g. materials handling, work-station arrangements, physical environment and work organization.

To facilitate the learning-by-doing with a view to self-help, good training tools should be developed and used. Particularly important are the action check-list, audio-visual aids illustrating local low-cost examples and small-group work. There is much scope for the improvement of small enterprises by facilitating self-help through the sharing of experience.

References

Bromley, P., 1985, *Planning for Small Enterprises in Third World Cities*, Oxford: Pergamon Press.

Dy, F. J., 1984, Improving working conditions and environment: problems and approaches in the Asian context, in *Humanizing Work*, pp. 74–91, Manila: Institute of Labor and Manpower Studies.

ILO, 1985, *Improvement of Working Conditions and Productivity in Small and Medium-sized Enterprises: Intensive Training Course and Short Training Course*, Madras, October 1985, Proceedings, Geneva: International Labour Office.

ILO, 1986, *Improvement of Working Conditions and Productivity in Small and Medium-sized Enterprises: Training Course*, Bangkok, 8–29 April 1986, Proceedings, Geneva: International Labour Office.

ILO, 1988, *Making the Enterprise More Efficient and a Better Place to Work: Training Course*, Noida and Ghaziabad, 16–20 November 1987, Proceedings, Geneva: International Labour Office.

Imada, A. S., 1985, Participatory ergonomics: its utility, its appeal and its necessity, in Brown, I. D., Goldsmith, R., Coombes, K. and Sinclair, M. A. (Eds) *Ergonomics International 85*, pp. 364–6, London: Taylor & Francis.

Kogi, K., 1985a, *Improving Working Conditions in Small Enterprises in Developing Asia*, Geneva: International Labour Office.

Kogi, K., 1985b, Participatory approach in applying ergonomics for workplace improvements in developing countries, in Brown, I. D., Goldsmith, R., Coombes, K. and Sinclair, M. A. (Eds) *Ergonomics International 85*, pp. 370–2, London: Taylor & Francis.

Kogi, K., 1987a, Group assessment for finding practical ergonomic improvement: experience in some developing countries, in *Ergonomics in Developing Countries: An International Symposium*, pp. 484–7, Geneva: International Labour Office.

Kogi, K., 1987b, Coping with industrialisation: experiences in small enterprises of Asia, in Suzuki, T. and Ohtsuka, R. (Eds) *Human Ecology of Health and Survival in Asia and the South Pacific*, pp. 213–7, Tokyo: University of Tokyo Press.

Kogi, K. and Sen, R. N., 1987, Third World ergonomics, *International Reviews of Ergonomics*, **1**, 77–118.

Louzine, A. E., 1982, Improving working conditions in small enterprises in developing countries, *International Labour Review*, **121**, 443–54.

Sen, R. N., 1984, Application of ergonomics to industrially developing countries, *Ergonomics*, **27**, 1021–32.

WHO, 1981, Education and training in occupational health, safety and ergonomics, Eighth Report of the Joint ILO/WHO Committee on Occupational Health, *Technical Report Series No. 663*, Geneva: WHO.

Wisner, A., 1985, *Quand Voyagent les Usines*, Paris: Syros.

Chapter 5

Design Decision Groups — A Participative Process for Developing Workplaces

John R. Wilson

Introduction

The use of participation and participative processes is lauded by many workers in ergonomics and job design. Gains are said to be considerable, both in direct contributions to the problem at hand and in systemic terms. Primarily, these gains are said to lead participants to feel that they 'own' any solution or design that results and thus to have a greater commitment to its success; also better information will come from those who know most about the existing system or jobs, which should lead to better solutions. The systemic spin-offs may include greater awareness of such ergonomics and job-design issues and of ways of tackling them, and an increased feeling of belonging within the organization (Wilson, 1991). Such motivations lie behind the participative work-design ventures of, for example, Boel *et al.*, (1985) Montreuil and Laville (1986), and Teiger and Laville (1987).

Of course, it must be accepted that participation, in whatever application, has its critics both as a concept and in the way it is applied (see Wilson, 1991). Nonetheless, the potential beneficial outcomes, evidence from the limited case literature available, the application-oriented nature of the approach, and the basic person/user oriented philosophy of participation all commend themselves to ergonomists.

This chapter looks at participative methodology in a particular application domain of ergonomics or human factors — workplace layout and design. In particular, a well-tested methodology, the use of design decision groups (DDGs), is described in concept and in practice.

Theoretical and practical background

In the late 1970s and early 1980s Denis O'Brien, an ergonomist at the UK Government Home Office, proposed a number of methods for running creative meetings and for involving users in systems design (e.g. O'Brien, 1981a,b). He took and adapted theories and techniques from market-research and design and from the literature on creativity and innovation, in particular using non-directive and group discussions and developing 'thinking tools'. Central to all the approaches he used is that 'current users can be encouraged to talk about, write down or draw attitudes, experiences, visions or opportunities which can be more relevant, perceptive, creative, and holistic than those achieved by traditional systems analysis techniques' (O'Brien, 1981a, p. 146). Although the approaches can be seen as risky, in that they contain emotionally charged input and also may put users at variance with management, the author nonetheless feels that better ideas or decisions can emerge in a shorter time-scale than with more conventional methods.

With such relatively free-form exercises, the design of the meeting or group session is of vital importance. To this end O'Brien describes a number of techniques or tools within what he calls shared-experience events (SEEs). Primary amongst these are the structuring of contributions from subgroups whilst others listen in silence, only later criticizing; use of a dynamic agenda, formed at first from a 'word map' exercise (see later), but flexible enough to change at need; use of drawing as far as possible, which enhances concept development, communication of ideas, and also helps in structuring the process; and, in common with many other seemingly casual techniques, very careful preplanning and preparation.

O'Brien's ideas were used in a study of Fire Brigade control room design; a design method was used which moved in a number of sessions from use of paper-and-pencil creative exercises through modelling to simulation trials (Langford, 1982). Subsequently, the present author has, with colleagues, adapted the method, called it 'design decision groups' and used it in a number of applications (Grey and Wilson, 1983; Grey et al., 1987; Wilson, 1988).

The process is described in the remainder of this chapter with reference to application in the design of library issue desks and, particularly, retail checkouts. The main differences between our work and that of Langford (1982) is that we worked to a shorter time scale, used as participants, representatives of these who might well one day use the new system or layout as against those who know that a new system will shortly be implemented at their workplace, and carried out our work in a neutral laboratory rather than at the user groups' places of work. Because of these differences, our users were paid volunteers.

In the case of library issue desks, we were interested in the layouts of furniture and equipment which were preferred by librarians and library assistants. In particular, we were interested in the participants' preferred layouts to cater for the introduction of scanning technology, but also sought design criteria relevant to all types of issue systems.

The technique was used with two mixed groups of library staff — one with experience of the particular issuing system and one without. Each group comprised eight paid volunteers, from different libraries, and covering a range of job functions. Both groups came to our laboratory for three, 3-h sessions, in successive weeks.

Our second study discussed here involved producing design concepts for a 'new' type of retail checkout, for a department-store-type environment. Purchases were predicted to be from two to eight per customer of greatly varying shapes, sizes and weights. A number of criteria were given: checkouts should be easy to manage for families; staff–customer communication, both formal and informal, should be made as pleasant as possible; children should not be or feel 'dwarfed' by surrounding fittings; and goods registering should be by hand (gun) scanner, carried out in the trolley or basket.

The DDG process in this case was to produce initial concepts for the design of such a checkout, particularly with a view to scanning feasibility, customer interaction, and till and keyboard use. Modifications to account for other restrictions (such as access, customer throughput, cabling and space utilization) were to be added later by the investigators.

Two groups each of six checkout staff were used, all paid volunteers, none of whom had any experience working with the particular scanning technology under investigation. In this case, two 4-h sessions were held with each group in successive weeks.

It should be emphasized again that in both cases the participants were representative users (as compared with user representatives), who may or may not use the new technology under investigation at some future time.

Design decision groups procedure

Scene setting

In order to enable (not force) people to participate in meetings, exercises, investigations, etc., certain general principles can be observed. They should feel that there is some purpose to participating, that outcomes will be of benefit to themselves and/or others. They should feel that they have enough information and knowledge to make a useful input, and should feel confident enough to ask for or to obtain such knowledge where they feel it is lacking. One important but tricky requirement is that the group must not be overtly directed or feel that this is the case; however, the group must have direction and be purposeful; disruptive or domineering members must be subtly reined in and not allowed to dominate meetings (see Wilson, 1991, (figure 2) for further requirements).

The above requirements for the DDG process can be achieved in several ways. We felt that holding meetings at an independent site (our laboratory) and emphasizing our independence and the confidentiality of the details of the process were important. Meetings were facilitated rather than run, generally with one investigator acting as introducer, prompter, interpreter and general resource, with a second investigator on hand as technical resource, including making and serving refreshments, distributing or showing graphical and visual aids, supplying materials and so on. The second investigator was also responsible for recording the proceedings, using video and photography (as unobtrusively as possible), and for note taking. All sessions were allowed to run in a relaxed manner, with everybody being gently encouraged to contribute, and with a very flexible agenda. At the same time, the investigators were well aware of certain tasks or exercises which had to be completed, and certain information which had to be derived, within particular time-scales. We have found that, although certain guidelines for doing this can be followed and can help, it is not surprising that it also takes a certain sort of person to facilitate the groups. Not everyone has the ability to communicate and subtly direct, without dictating or dominating.

All potential participants were contacted at their place of work and informed that they would take part in confidential discussions about the design of their workplace, how much they would be paid, and the dates and times when we would need them. Library staff largely came for evening sessions; retail staff came on their afternoons off. All participants were collected and returned by taxi. Payment, which compared very favourably with their normal wages, was made in one lump sum at the end of all sessions.

In the remainder of this section the stages of the DDG process are described and illustrated.

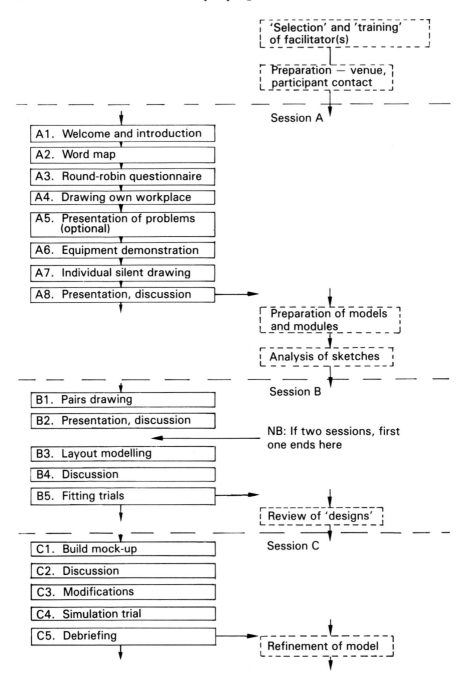

Figure 5.1. Summary of the design decision group process.

This is shown in figure 5.1 as a three-session process; however, if time or other resource restrictions dictate, two, somewhat longer, sessions can be used, as in the retail case.

Session A

Introduction

This includes a welcome, introductions, an explanation of who the design meeting organizers are and a general overview of what is expected of the design sessions. Refreshments are served and every effort is made to establish a comfortable, relaxed environment.

Word map

'Word maps' or 'visual maps' are constructed by every participant, volunteering words, which they feel are connected with the particular design topic for a 5–10 min period. O'Brien (1980, 1981a) sees the technique as defining the total problem space and helping produce the dynamic agenda. In our case we simply asked participants to make an inventory of objects and equipment required at their workplace. This acts as an 'ice-breaking' device and also participation is such an easy, non-stressful exercise it breaks down any inhibitions that quieter members of the group might have about expressing their ideas in front of their colleagues. Participation was encouraged by going 'round the table' several times (see figure 5.2).

Round-robin questionnaire

Participants are presented with a series of very simple open-ended questions such as: 'A good work counter is . . .?' or 'Problems at my workplace are . . .?'. Each question is printed on a separate sheet, and these are passed around all participants with each attempting to complete the sentence with a different ending to that chosen by the others (figure 5.3). This method is preferred to the alternative of each participant filling out an individual questionnaire for two reasons. Firstly, it makes the exercise a little more productive, in that each participant has to think of a different ending from those given by the others, thus providing more information by reducing duplication. Secondly, it can act as a stimulus. If, for example, participants had missed the point of the exercise, or were unable to think of any ideas for a certain sentence, then seeing what had been written by others may trigger a line of thought that could otherwise have lain dormant.

Drawing of own workplace

As a further ice-breaking exercise, as preparation for the more important drawing exercises to come, and to provide investigators with extra information, participants were asked to draw their existing workplace, indicating its good and bad points.

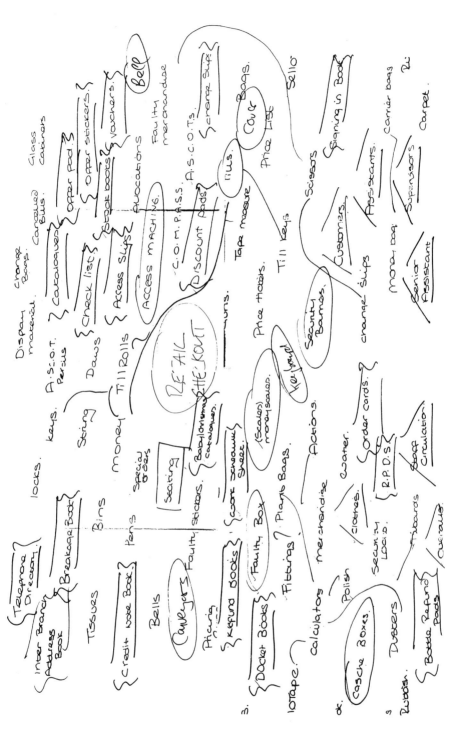

Figure 5.2. Part of a simple word map.

Seeing Rubbish and not being able to find anything but piles of faulty goods stuck up in one corner

When I'm working, the things I dislike in a checkout are. *having*

to stand and look down on customers, not having enough room to write and bag items.

not having the right equipment at hand to deal with customers paying by ~~cheque~~ *credit cards.*

Having to stand, waiting for someone to answer the bell.

To Show a customer where something is. (point).

Having papers etc hanging around.

Merchandise hanging around under your feet.

Having catalogues on the top near the tills.

Having no spare till Rolls.

*People eg Sales Assistant: not coming on time when sound ...
Be ... and not answering bell*

*when you can't find price list or a price in book.
when someone wants a item checking over or looking at.
poor Bag on Floor*

*Not Having enough Time To Talk To Customers
when there are no customers to serve. but you can't come off till another person comes on.
when you can't find anything when there are so many draws.*

Figure 5.3. Example of a round-robin questionnaire.

Slide presentation of ergonomics problems

Slides are selected from associated field-work, covering a range of topics: different types of layout, different types of equipment, etc., and, most importantly, a series showing how these pieces of equipment have been poorly installed in certain environments. The slides emphasize such features as poor seat height, lack of leg room, poor work-surface height, lack of work space and bad positioning of equipment and displays.

This stage is intended to provide a logical progression from the round-robin questionnaire. At that stage, the participants are asked to think about what makes good and bad designs, but only have their own experience to draw upon. The slide show is intended to make participants aware of systems and layouts that they may not have encountered before, and also to emphasize problems that they may not have considered whilst being influenced so heavily by their current system. Slides of a variety of different workplaces and systems can

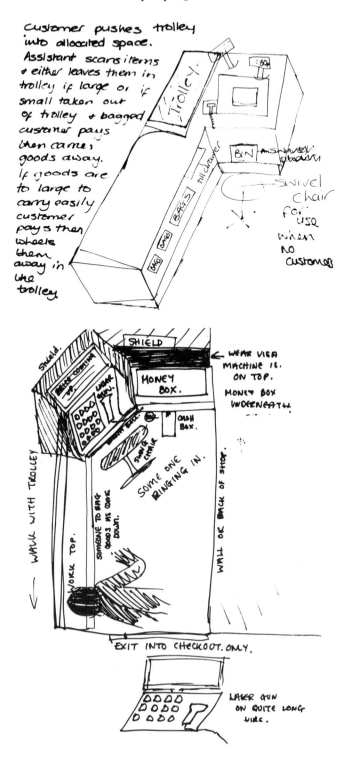

Customer pushes trolley into allocated space. Assistant scans items & either leaves them in trolley if large or if small taken out of trolley & bagged customer pays then carries goods away. If goods are to large to carry easily customer pays then wheels them away in the trolley.

Trolley.

BIN

swivel chair for use when no customers

SHIELD

← WEAR VISA MACHINE IS. ON TOP.

MONEY BOX.

MONEY BOX UNDERNEATH.

CASH BOX.

SOME ONE RINGING IN.

← WALL WITH TROLLEY

WORK TOP.

SOMEONE TO BAG GOODS AS COME DOWN.

WALL OR BACK OF SHOP.

EXIT INTO CHECKOUT. ONLY.

LASER GUN ON QUITE LONG WIRE.

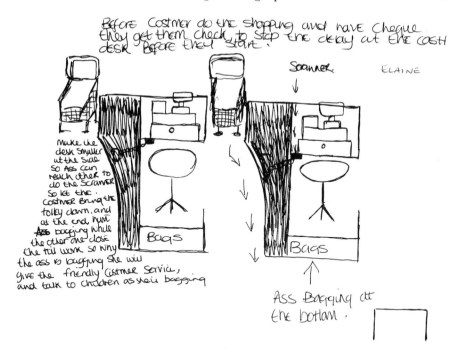

Figure 5.4. Three examples of individually drawn proposed work spaces, showing individual differences in design ideas and how these can be illustrated and communicated. Considerable colour coding is used in the originals.

also emphasize the point that we are required to design something entirely new, and that suggested designs should be unimpeded by a narrow range of experience.

Presenting slides in this way was not done by Langford (1982). We are conscious that such a procedure must not direct our participation in any particular direction, either of criticism of existing systems or of ideas for new ones. It may not be appropriate in all circumstances.

Equipment demonstration

When systems or equipment are to be installed which may be new to any or all participants it is advisable to demonstrate these and their operation. Actual examples should be used where possible; if not illustrations could be used.

Individual silent drawing

Each participant is asked to produce sketches on A4-sized paper of both plan and perspective views of their ideas for a good new workplace, incorporating any new systems. These should detail proposed layouts for people, work counters, modules and equipment (figure 5.4). At this stage, only very general ideas need to be produced, and participants are discouraged from becoming too involved in detail (e.g. precise measurements of work surfaces, or exact location of documents). To call this stage a 'silent drawing exercise' is perhaps overstating the point, but it is beneficial to discourage communication in the hope that a variety of different ideas and designs will be produced for use in the later discussion stage. Each individual is encouraged to try out several rough ideas before producing a final sketch suitable for presentation. Different coloured pens are available for colour coding various items of equipment or furniture.

Individual presentation and discussion

Each participant is asked to present their final sketch to the rest of the group and to explain the reasoning behind the design. The designs are discussed and gently (the role of the facilitator) criticized.

After each of the sketches has been discussed many problems should be highlighted, and certain designs may be seen as 'superior'. The role of the facilitator is then to encourage a general discussion, concentrating on one or two particular designs and bringing up any points that may not have been previously considered, probably by referring back to issues arising from the word-map and round-robin stages.

Session B

Drawing in pairs

At the beginning of the session the facilitator briefly runs through some of the points arising from the sketches made in session A. The group is then divided into pairs or threes, according to the similarity of their ideas, and asked to produce a larger sketch, on A3 paper, of the plan for the workplace. This time more emphasis is placed on detail. Positioning of equipment is to be specified more precisely, as are seating and furniture design (figure 5.5).

Presentation and discussion

Each drawing is presented to the rest of the group by its authors. Discussion at this point is geared toward choosing the 'best' drawing, but no firm decisions need to be reached as each group will get a further chance to demonstrate their ideas using models of the system.

Simple layout modelling

Participants are shown empty spaces of approximately the size available at the future workplace site. Very strong corrugated cardboard models of equipment are provided, in addition to a supply of mock-up counters, tables, desks, consoles and other modules. A variety of actual seats, tables and benches is also provided, as is a stock of string, flat card, modelling knives and heavy tape. Although the use of card models may seem, at first glance, to be amateurish, there are several compelling advantages. Models and mock-ups are more portable and participants are not frightened of breakage so building model work stations becomes a fast, rapidly changing exercise. Extra pieces of equipment or new pieces can be found or created very quickly. Work-station, work-counter and module sizes and

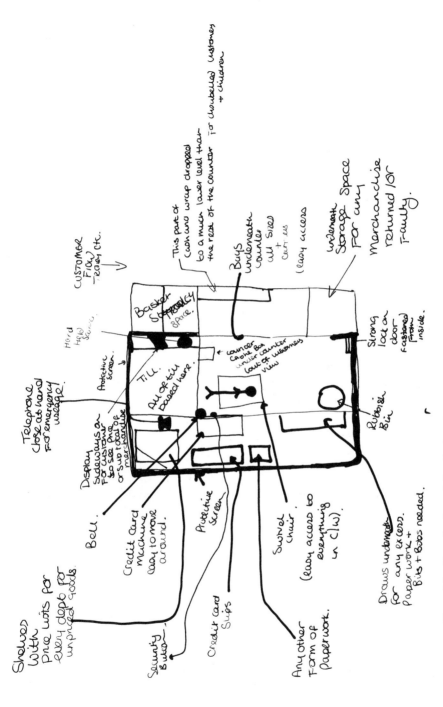

Figure 5.5. Example of a workplace developed by a pair of participants, and used as the basis for a model-building exercise.

positioning can be changed rapidly and easily. We have not had any participants who felt that they were taking part in a worthless or 'cheap' game.

Participants, still in subgroups of two or three, build a three-dimensional full-size model of their idea of the optimum workplace (figures 5.6 and 5.7).

Figure 5.6.   Initial planning for layout: two groups.

Figure 5.7.   Discussion of the juxtaposition of equipment.

Discussion

Discussion involving all participants is firmly geared toward eliminating any inferior design(s) and forming a consensus of opinion as to the optimal layout. This is often an amalgam of the various ideas represented by the different mock-ups. Limited 'walk throughs', testing aspects of the workplace, may be undertaken at this stage.

Fitting trials

On reaching this stage, a general workplace layout and the positioning of all the equipment should largely have been agreed. The next crucial factor to be tackled is that of work-surface heights and depths. These are determined through fitting trials, with the aid of adjustable metal stands holding the cardboard models, using all group members as subjects. In addition, however, the facilitator must ensure, at the time or later, that anthropometric characteristics of other potential users are taken into account.

Session C

Building a mock-up

On returning for the third session, participants build a complete mock-up (or two if there is still internal disagreement) of the preferred work station, integrating ideas on the placement and juxtaposition of equipment with work-surface height and depth data (figure 5.8).

Figure 5.8. Building a mock-up workplace.

Discussion

The discussion entails detailed consideration of the operational requirements at each particular work station; if two designs have been modelled, each subgroup must defend its ideas against those of the other group.

Modifications

As a result of the discussion, any modifications to the general structure of the work stations are made. If more than one design is still being considered, the best alternative is generally selected at this stage.

Simulation trial and assessment

The trials are aimed at simulating as many as possible of the scenarios, operations and tasks which are or might be undertaken in actual use. Participants take turns in playing the roles of staff and customers (figure 5.9), including children (figure 5.10). Final adjustments to the workplace model are made at this point.

Figure 5.9.   A simulation trial.

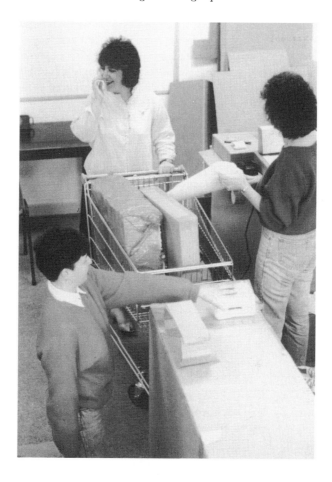

Figure 5.10. A trial showing (initially sceptical!) participant simulating a child.

Debriefing

Debriefing takes place after each session but this is the most important time of all. Participants are encouraged to discuss the value of the exercise and the perceived quality of their designs.

Refinements are made at the end of each stage, with the investigators accounting for constraints, equipment or tasks which may have been omitted (deliberately or accidentally) in the DDG sessions, or which may have come to attention later.

Design decision group outcomes

The outcomes of a DDG study, or other similar work, are varied. At the very least we have managed to find out, from the people who should know most about it, the problems or benefits associated with existing systems, the potential problems with the new system, and a number of concepts to be fed into the development process. At best, one or more feasible design alternatives for the whole work station is provided; this is an invaluable outcome, regardless of whether these designs are 'new' to the investigators or are combinations of ideas already developed. The process and setting also allow flexible and iterative testing of design concepts, whether derived beforehand or during the process.

It has been noticed in all our work with DDGs that the creative, yet relaxed, atmosphere and the resultant concentration of minds seems to encourage the expression of opinion or facts about other aspects of work redesign. During the libraries study especially we learned much of interest about training and support, feedback and other communication, autonomy, role clarity and other job-content and organization issues. This allowed us to make a more informed analysis of current situations, and to make better suggested changes for improvement. Although not used by us in such a context, it is suggested that the spin-off benefits will be considerable when such an exercise is conducted at a local level with people whose own technology, workplaces and jobs are to be changed. Awareness of what should be done to improve jobs and workplaces and of the criteria which should be applied, and the confidence to promote their own ideas can all help to create a more committed work-force, as well as healthier, safer, more effective and more satisfying jobs. The use of DDGs can thus give rise to direct gains through its outcomes in terms of design ideas, and can also give rise to systemic gains through the very application of the process.

References

Boel, M., Daniellou, F., Desmores, E. and Teiger, C., 1985, Real work analysis and workers' involvement, in *Ergonomics International 85, Proceedings of the 9th Congress of the International Ergonomics Association*, Bournemouth, September, pp. 235–7.

Grey, S. M. and Wilson, J. R., 1983, 'The ergonomics of library issue desks', unpublished report to the British Library, Department of Engineering Production, University of Birmingham.

Grey, S. M., Norris, B. J. and Wilson, J. R., 1987, *Ergonomics in the Electronic Retail Environment*, Slough: ICL (UK) Ltd..

Langford, J. B., 1982, *West Sussex Fire Brigade: Control Room Design*, London: Scientific Research and Development Branch, Home Office.

Montreuil, S. and Laville, A., 1986, Cooperation between ergonomists and workers in the study of posture in order to modify work conditions, in Corlett, E. N., Wilson, J. R. and Manenica, I. (Eds) *The Ergonomics of Working Postures*, pp. 293–304, London: Taylor & Francis.

O'Brien, D. D., 1980, New products: harmonious designs come from well-managed and coordinated teams, *Planned Innovation*, **3**, 111–4.

O'Brien, D. D., 1981a, Designing systems for new users, *Design Studies*, **2**, 139–50.

O'Brien, D. D., 1981b, 'Methods to enrich human communication and creativity', unpublished paper from UK Home Office, May.

Tieger, C. and Laville, A., 1987, 'How ergonomic methods can be used by non-ergonomists', presentation at 2nd International Occupational Ergonomics Symposium, Applied Methods in Ergonomics, Zadar, Yugoslavia, April.

Wilson, J. R., 1988, Communications issues in occupational ergonomics, in *Proceedings of the International Conference on Ergonomics, Occupational Safety and Health, and the Environment*, Beijing, China, October, pp. 13–19.

Wilson, J. R., 1991, Participation — a framework and a foundation for ergonomics? *Journal of Occupational Psychology*, **64**, in press.

Chapter 6
Participatory Ergonomics in two US Automotive Plants

J. K. Liker, B. S. Joseph and S. S. Ulin

Introduction

Background

In the 1970s and 1980s nothing less than a revolution occurred in the history of the US automotive industry. In this period of increased fuel costs and intense international competition, automotive manufacturers and suppliers attempted to transform their operations and corporate culture. It is commonly agreed that their very survival was at stake. The goal was to increase product quality, reduce product costs, design more desirable vehicles, and encourage employee involvement throughout the company, including on the shop-floor. Among the ways this is being accomplished are new quality programmes, new employee-involvement programmes, large expenditures on training facilities and trainers, large capital expenditures on new technology, shutting down many outdated and unproductive production facilities and opening new ones, and large cutbacks in white-collar and blue-collar staffing. These massive changes have touched every part of the industry and meant that the most progressive organizations which have kept abreast of the changes are in a constant state of flux and typically overloaded with new programmes. In the 1980s, a new addition to these programmes was the use of 'ergonomics' to improve the design of manufacturing jobs.

In the past, human-factors specialists were employed at the staff level but their contributions were limited mainly to vehicle design. The term 'ergonomics' in the automotive context became a buzzword to advertize that the vehicle was designed to be easy for the driver to operate and maintain. In the meantime, machine operators in plants producing the vehicle were slowly damaging their bodies as they performed poorly designed, repetitive jobs. Traditional work methods were used to design the jobs so that they could be performed within the performance standard set for the job; however, these were often not the best designs for the health and safety of workers. Particularly neglected have been cumulative trauma disorders; that is, a gradual breakdown of the muscles and joints because of repeated forceful exertions, awkward postures, and other related stresses (Armstrong *et al.*, 1986).

More recently, the three largest US manufacturers — General Motors, Ford and Chrysler — in collaboration with the United Automotive Workers Union (UAW) have all made some investments in ergonomics in their manufacturing operations. This has largely been through outside contracts for training and consulting services with university faculty and staff and through individual plant efforts. While there have been some efforts to apply ergonomics in the original design of new manufacturing facilities, most of the efforts have been aimed at redesigning existing operations. The most organized efforts have been in a small number of plants selected as 'pilots' with the hope that by creating a few successful models of change, other plants can learn from these early experiments.

This chapter describes and analyses pilot programmes in two different automotive parts plants that are part of the same major automotive company. Both pilot programmes were set up specifically to test the benefits of including worker involvement in ergonomics. The term 'participatory ergonomics' as used in this book takes on a very broad meaning encompassing employee participation in changes in the work environment to foster physical and psychological health. Both of the participatory ergonomics programmes described in this chapter had a more specialized focus, involving worker participation in the design or redesign of repetitive, manual jobs in order to minimize cumulative trauma disorders.

Why participatory ergonomics?

The problem of implementing ergonomics

Epidemiological studies from many industrial sites have indicated that musculoskeletal injuries in the workplace known to be associated with poor design of jobs are a major cause of lost work time and high worker-compensation costs (Kelsey *et al.*, 1978; Joseph, 1986). Despite years of ergonomics research and a plethora of data, it would seem that the use of ergonomics has not made adequate inroads into the workplace. There are clearly many reasons for this absence, many peculiar to particular plants. However, there are several generic organizational factors that inhibit the application of ergonomics.

Liker *et al.* (1984) proposed the presence of four obstacles to the utilization of ergonomics in the workplace:

1. *A lack of general ergonomic knowledge.* Plant personnel generally lack an understanding of the basic principles of ergonomics including human capabilities and limitations, physiology, psychology, manufacturing processes, machine design and process design. While individual specialists may have a limited part of this requisite knowledge, no one person will have all the necessary general knowledge.
2. *A lack of specific job knowledge.* This knowledge comes from experience with the particular machine, experience that workplace designers typically do not have.
3. *Poor interdepartmental communication.* A key to successfully applying ergonomics is communication between those departments in the plant that collectively have much of the necessary knowledge and political leverage for successful application of ergonomics.
4. *Perceived cost benefits (subunit interest or suboptimization).* People tend to act in their own self-interest. Plant interests tend to have a short-term focus, reducing the incentives for

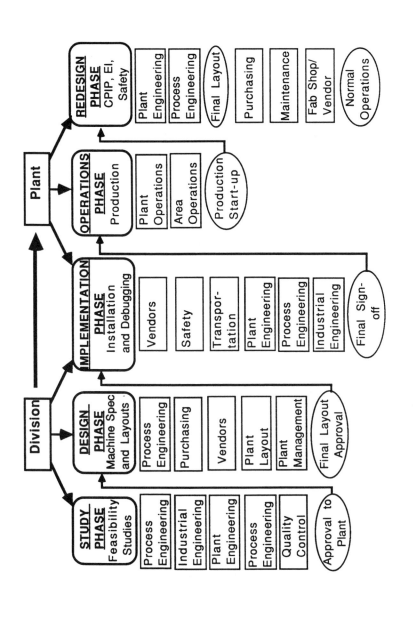

Figure 6.1. Organizational process for the installation of new and the maintenance of existing equipment.

conducting ergonomics studies as their results (e.g. health effects) are often seen only in the long term. Perhaps for this reason, human-factors specialists, if they exist at all in the company, have relatively little power compared with production staff (Perrow, 1983).

These obstacles come about, in part, from the organizational complexity of large, bureaucratic corporations. Figure 6.1 illustrates the complexity of the organizational process required for the installation of new and the maintenance of existing equipment within a division of the automotive manufacturer on which this chapter focuses. In this division, phases differentiate periods of time in which groups of units have to complete a task before going to the next phase. First, new processes are studied (study phase) and designed (design phase) at division engineering. Little plant input is solicited in this phase. Next, division and plant engineering together install and debug machinery (implementation phase). This procedure involves a complex series of actions whereby process plans are sent to selected vendors, interpreted and built to specifications, delivered, and installed in the plant using resources from the plant, vendor and division. Unless the plant is willing to bear costly delays and excessive expenditures, few changes can be made on machinery, for ergonomic or any other reason, during this phase. Consequently, the plant must wait until the machines are delivered and operating under those specifications (known as final sign-off) before changes can be made. After debugging, normal operation and maintenance of production proceeds (operation phase). However, often there is a need for process improvements or other redesigns to update equipment (redesign phase). The redesign phase is where ergonomic adjustments are typically made, generally only when a clear problem arises (e.g. workers suffering from carpal tunnel syndrome). Depending on the cost, the plant usually controls

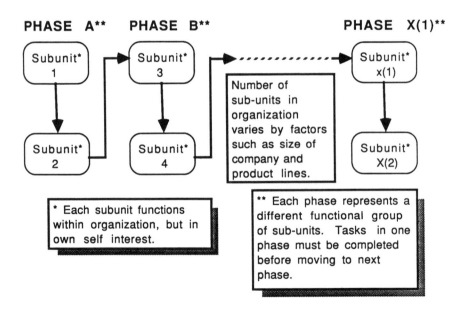

Figure 6.2. Simplified model of organizational politics.

redesign activities. However, due to limitations on cash and manpower resources, lost production from shutting down the machines, and other plant priorities, this activity is often limited and delayed until the shutdown will cause the least cost.

If one examines this process closely, it bears a striking resemblance to the organizational politics model depicted in figure 6.2 (Allison, 1971). This model represents large organizations as bureaucracies composed of individual organizational units that are separated into functional phases. These units act in their own self-interest to achieve their desired goals. Interdependent units with different goals or perceptions of how to reach a common goal or who are jointly dependent on scarce resources are likely to experience conflict between them. Therefore, the workplace-design process shown in figure 6.1, which involves interdependence of so many parties, is a likely breeding ground for conflicting goals and intergroup tensions. Moreover, the fact that parties who make key decisions on manufacturing processes are separated geographically and organizationally creates a high probability of communication breakdown (Allen, 1977). Given the nature of this bureaucratic process, it is not surprising that workplaces are often not well designed ergonomically.

The solution

If we assume that effectively applying ergonomics requires the cooperation and communication of a variety of groups of people who do not normally talk to each other, the obvious solution is to bring these people together in meetings. One such approach is the use of task forces trained and established specifically to use ergonomics tools and knowledge to improve jobs. This approach has been strongly advocated by Pope (1987, p. 454): 'An ergonomic approach, soundly based on biomechanical principles, will be effective in reducing such injuries if the correct management approach is taken. The team should employ representatives from all relevant constituencies and should develop programmes for hazard identification, prevention, implementation, measurement and training'.

In reality, successfully managing the process of change requires more than simply getting people together in meetings. First, many programme-design decisions must be made early on, such as: Who should be on the task force? How many task forces are needed and what areas of the plant do they cover? What training should task-force members receive? Who else beyond task-force members should be trained? How should both shifts in a two-shift operation be involved? Second, after the initial start up of the programme it must be carefully managed and supported. For example, meetings must be effectively run, individual members must work together as a team, organizational politics must be carefully considered and managed, technical expertise in ergonomics must be developed, and the resources needed to change jobs must be provided.

This chapter illustrates how administrative units within two automotive parts plants addressed the challenges of starting up and maintaining ergonomics programmes. They did so in different ways and with different degrees of success. As much as possible we attempt to elucidate the differences in the ways ergonomics was organized and practised between areas within plants, and between plants, and draw out implications for effectively managing participatory ergonomics programmes. We, of course, acknowledge the limitations of drawing strong scientific conclusions based on a small number of case studies. Nonetheless, the detailed case descriptions that follow yield insights into the challenges of participatory ergonomics and provide exposure to alternatives available for meeting these challenges.

Case background

The plants

Two plants set up participatory ergonomics programmes under the guidance of the authors over a 4-year period. We briefly describe the plants in this section.

The chassis plant

This was our first attempt to study the effectiveness of a participative approach to ergonomics. The chassis-components plant is a captive automotive supplier for a major automotive concern in the USA. Located in Michigan and in operation since 1968, it occupies over 1.75 million square feet; 1.5 million square feet of floor space is devoted to the manufacturing processes necessary to make chassis components. Some of the operations involve machining on large transfer lines, with some assembly, while other parts are machined using several different processes. As of November 1985, the number of employees at the plant was 1697. The plant experiences very little monthly fluctuation in employment.

During the period between 1980 and 1982, a severe recession occurred in the American automotive industry. Companies consolidated operations to reduce overheads and to increase efficiency, including closing unprofitable plants and operations. The chassis plant was identified to be closed unless management and labour together could come up with a plan to make the plant competitive. In response to this ultimatum, several things happened to save the plant. Two of the most important were the streamlining of plant management complexity by reorganization into several autonomous areas each governed by an area manager (as described below) and a change in management philosophy toward people-oriented projects, including an active programme of employee-participation groups (EPG) who meet weekly, labour–management team meetings, and the eventual acceptance of the ergonomics programme into the plant. Product cost and quality improved dramatically within a short time of introducing these human-resource innovations. Plant management attribute their improved performance to their new approach to human-resource management and the plant became one of the corporation's models in this regard.

The chassis plant ergonomics programme officially began late in December 1983. Researchers from the University of Michigan had received a grant from an automotive industry interest group to study ergonomics, and this plant agreed to be a study plant for that purpose. The proposal for implementing a participatory ergonomics programme was made at a meeting of all managers and upper-level supervision from the production areas and leaders of the UAW at the plant level. The progressive managers of this plant agreed to work with the university on the programme and, in particular, agreed to involve hourly employees. For its part, the university agreed to provide training and ongoing technical assistance by dedicating a graduate student in ergonomics quarter-time to the plant. (A detailed account of the chassis plant case can be found in Joseph (1986).)

The stamping plant

Most of the metal parts of an automobile begin at a stamping plant. The process begins with large rolls of steel and involves blanking, stamping, and assembling the parts. The stamping plant that developed a participatory ergonomics programme is part of a different division of the same major automotive manufacturer as the chassis plant. The stamping plant produces such products as fenders, aprons, wheelhouses, cowl tops, quarter panels, and many other products. Approximately 80 per cent of the parts they build are for older, established car lines and approximately 20 per cent are for new car lines which were introduced at the time of the ergonomics programme.

Ground for the stamping plant was broken in 1936 and production began in 1939. It is the oldest stamping plant in operation within the company and much of its technology is outdated. Consequently, there have been rumours of the plant closing down completely for over 20 years as more modern stamping plants have been constructed at other locations. Within the five years prior to the study, the plant had been continually reducing the number of hourly and salaried employees.

From the beginning, the plant manager warned the researchers who were proposing a participatory ergonomics programme that the plant may close and that there would be continual downsizing. There were 2200 hourly and 325 salaried employees working at the plant at the beginning of the ergonomics programme. By the time the participatory ergonomics programme had been in operation for almost 2 years, hourly employment was reduced to 1500 and another substantial reduction in work force was being planned.

This is not to portray the stamping plant as an old, decaying plant devoid of energy and innovative spirit. Like the chassis plant, their innovative use of new management approaches had made them a model plant in their company. Despite their old facilities and equipment, in 1986 they were the best quality stamping plant within the company. Moreover, their costs were competitive with those of newer stamping plants having more modern and highly automated facilities. In 1986 they were second in costs relative to the budget of all plants in their division, which includes relatively affluent assembly plants.

The stamping plant ergonomics programme was officially launched in June 1985. The automotive division of which this plant was a part had funded a multi-year university programme of basic research, training, and consultation in ergonomics. As part of the broader grant, the stamping plant subproject focused on researching new ways of organizing ergonomics programmes in manufacturing plants. The stamping plant was selected as the study plant because of the plant manager's enthusiasm for ergonomics and the plant's reputation for innovative human-resource management programmes. A proposal was made at a meeting of top managers and UAW officials at the stamping plant who agreed to work with the university on the programme and who made a commitment to involve hourly employees. As in the chassis plant, the university made a commitment to provide training and a quarter-time graduate student in ergonomics. However, in this case, there was also a broader commitment at divisional level to support ergonomics efforts in plants.

Comparison of the stamping and chassis plants

In many ways the two pilot plants were very similar. Among the *similarities*, both were:

1. automotive-parts plants;
2. part of the same major automotive manufacturer;
3. comparable in size (about 2000 employees);
4. old plants with a recent history of shut-down threats and reduction in labour-force size;
5. outstanding in their innovative approach to human-resource management (participative management, cooperative union–management relations, and autonomous area management);
6. performing competitively on cost and quality compared with newer plants having more modern facilities, which was attributed to their effective human-resource management;
7. running on a very tight budget; and
8. running with a very lean staff, including production personnel and support staff.

At the time when the ergonomics programme was initiated, both plants had been recently reorganized based on an area-management concept. The organization of each plant includes two production areas (A and B) governed by an area manager who reports to the plant manager. Under each area manager is a suborganization that controls daily activities. This consists of superintendents for each production department that manufactures or assembles related components and a set of support functions assigned specifically to the area (e.g. maintenance and engineering). Under each superintendent, supervisors are responsible for overseeing the individual production departments. The supervisors are in direct contact with the workers. At this level, the organization is quite traditional: normal lines of communication between workers and managers require the workers first to contact their supervisors, although the plants were beginning to experiment with self-managing teams and eliminating the traditional supervisory role. There continued to be some centralized support activities in each plant (e.g. personnel, labour relations, accounting, and quality control), but to a large degree each production area operated autonomously. For this reason the two areas were sometimes referred to as two plants under one roof. Therefore, the ergonomics programme operated relatively separately in each area and we were able to compare two case examples in each of the two plants.

In other respects the two plants were very different. First, the core manufacturing operations in each plant were different. One plant was a stamping plant which stamped out parts from large rolls of sheet metal and welded parts together. The other was a chassis plant in which parts were machined on large transfer lines and some assembly was required. Among the many differences in these technologies, the jobs in the stamping plant were relatively homogeneous which meant workers who knew one job could also contribute to discussions of many other jobs in the plant. This was not the case in the chassis plant where jobs were relatively unique and, therefore, a worker's knowledge was very specialized to his or her own job. Second, the degree of division-level support for ergonomics was much higher in the stamping plant's division. The stamping plant was part of the same division as assembly plants and the top management of this division had contracted with the university to help bring ergonomics into plants. There was considerable funding for training as well as division management pressure on plants to use ergonomics to improve their jobs. By contrast, in the chassis plant only modest levels of funding were available from an outside industry interest group and there was no division-level pressure for ergonomics — they had to be self-motivated.

Programme development and outcomes by plant

The programmes in both plants can be viewed as progressing through three stages. The first stage, *laying the groundwork*, is a preparatory phase. Laying the groundwork means selling ergonomics to many people at the plant, including management, the union, support staff, and hourly workers who will be asked to participate in the project. It also means holding meetings to design the organization, select the participants, develop training schedules, and identify the mission of the programme and the microstructure and roles of the committees in the programme. The second stage, *getting our feet on the ground*, starts when the ergonomics task forces actually begin to meet. At this point, these groups of individuals must learn to become a team and work together as a cooperative unit and also must learn how to apply ergonomics tools to the analysis of jobs. The third stage, *up and running*, is a period in which the programme has achieved a relatively steady state. In this stage a major challenge is to keep participants' motivation and commitment high.

We use this three-stage model below as an organizing framework for describing the development of the participatory ergonomics programmes in the two automotive parts plants. This is not to suggest that any programme exactly fits this model. For example, it is debatable exactly when one stage ends and another begins, particularly between stages two and three. However, each stage poses a distinct set of challenges that must be met and the stage framework provides a way of looking at the process by which participatory ergonomics programmes are developed and maintained.

The chassis plant ergonomics programme

The history of the ergonomics programme at the chassis plant is summarized in figure 6.3. As can be seen, the programme was set up differently in the two main production areas of the plant. Much of the analysis below focuses on these area differences.

Stage 1 — Laying the groundwork

After the initial meeting with the plant management and union leaders at which the programme was proposed and accepted, a second meeting was held to design the organizational structure of the programme. The result was a programme design that called for a plant advisory committee and separate ergonomics committees in each of the two production areas of the plant. Based on this design two levels of ergonomics training were scheduled. The more extensive 8-h course was offered to engineers, superintendents, and maintenance personnel. A shortened 4-h course was offered to plant management, supervisors, union officials and hourly employees. Hourly employees were included in the shorter course to keep their time away from production to a minimum.

The programme organization was modified over time. Figure 6.4 shows the ergonomics organization used in the case study for most of the data-collection period. There were four groups: an advisory committee, a task force from area A, an area A carrier-machining-worker group, and an area B spindle combined task force/worker group. The responsibilities of each group are described below.

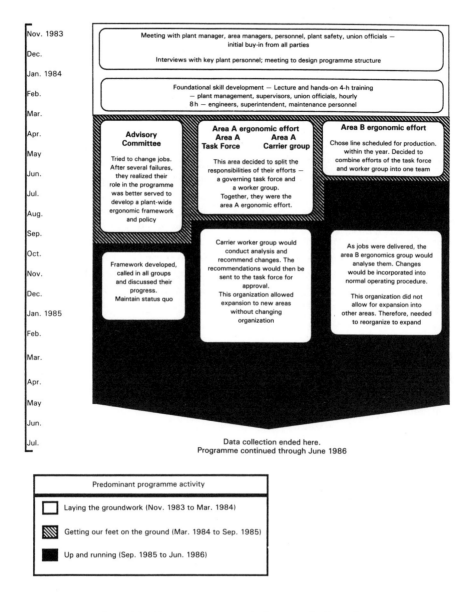

Figure 6.3.   Ergonomics programme development at the chassis plant.

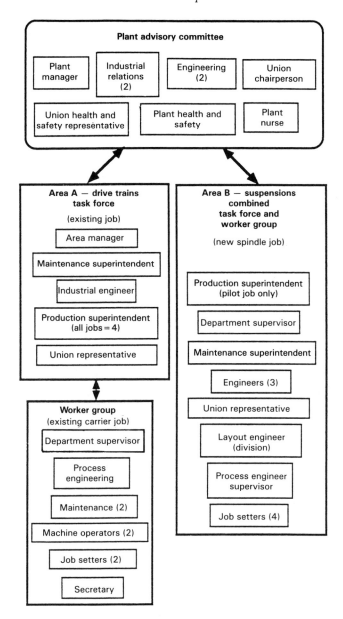

Figure 6.4. Final chassis ergonomics organization.

Advisory committee

This committee consisted of managers, union officials, and support personnel responsible for the entire manufacturing facility; it overlapped considerably with the plant operating committee. The group's major task was to set and continually monitor all goals and policies of the ergonomics programme. Additional tasks included out-reach activities designed to advertize major accomplishments inside and outside the company and support for those projects that incurred high capital or personnel costs (dollar value in excess of $10 000). Most members received 4 h of training.

Area A task force
Area A decided to organize a two-level system: a task force representing the entire area and a worker group focused on one particular line. In this way as the task force shifted to other lines, they could form new worker groups from the workers on those lines. The task force's major responsibilities were to:

1. guide subordinate worker groups to priority jobs needing change;
2. act as a clearing house for recommendations of job changes and suggestions from worker groups; and
3. provide resources, including funding and technical support, to the worker groups and the general programme.

This group, consisting mainly of middle managers, went through several reorganizations during the data-collection period. However, a certain core group remained intact throughout the study. Three of its eight members were not trained in the original ergonomics course offered at the plant. However, they were trained in specific analysis tools as the programme progressed.

Area A carrier-machining-worker group
The steering committee formed this group to look at an existing job in an area with traditionally high absenteeism and poor labour relations. Members felt that these problems existed partially because of the physically demanding efforts required by the excessively heavy parts. The carrier-machine jobs, which involve the handling of raw castings weighing between 35–45 pounds on a machining line, had been running for several years. The group's responsibilities included:

1. identification location of physical stresses on operators performing the jobs;
2. brainstorming ways to reduce the stress; and
3. selecting the best solutions on the basis of ergonomic considerations, costs, efficiency, production schedules, and feasibility.

The group members were primarily hourly employees with some middle managers (including the maintenance supervisor). Three members of the group received no training. In addition, one machine operator left the group after approximately eight meetings.

Area B spindle combined task force/worker group
Area B chose to direct ergonomic efforts towards a new 'spindle' line that was in the process of installation — all operations were part of the manufacturing process for a single spindle. This line was chosen for two reasons: first, this would be a proactive design process incorporating ergonomics in the early design phase; and second, making changes to existing jobs is a drain on area resources. However, as the spindle line was a new process, the division responsible for installing the equipment was also responsible for any changes to the

equipment until final sign-off at the plant. Therefore, the use of division resources in conjunction with plant personnel to make the changes would minimize the costs charged to area B.

The group's responsibilities included all those outlined for the area A task force and carrier worker group combined, including identifying problem jobs, brainstorming control measures, and selecting and implementing the best solutions. As can be seen in figure 6.4, this combined task force/worker group was large (13 people) and consisted mainly of middle managers with a large minority of hourly employees. This group was similar in composition and responsibilities to the ergonomics 'operating committees' used in the stamping plant (discussed below).

Summary of organization in areas A and B

Because area B combined the task force and the worker group into one resource group, it bore the responsibility for both designing *and* implementing changes. In comparison, in area A the task force was responsible for implementing the changes, allowing the carrier worker group more time to analyse additional jobs. With the exception of the advisory committee, all committees met weekly on company time. The graduate student in ergonomics attended all committee meetings, and provided technical support to analyse jobs and develop ergonomic solutions.

Stages 2 and 3 — Programme development and maintenance

In this section we discuss both stages 2 and 3 — programme development and maintenance, referred to above as 'getting our feet on the ground' and 'up and running', respectively. A summary of these phases is provided in figure 6.3.

One way of representing the differences in the committees during these phases is to consider how the committees spent their time. During the course of the investigation, estimates were made of the percentage of time devoted to five generic activities for each meeting. These measures included the percentage of time spent on programme maintenance, problem identification, workplace design, implementation, and non-ergonomic activities. The results are summarized in figure 6.5 for meetings held in stages 2 and 3.

Programme maintenance included those administrative activities necessary both to start the group running during the development phase and to keep it running smoothly during the programme maintenance phase. Examples are the development of action plans, discussions of how to increase participation of quiet members, and discussions of how to increase member attendance. By contrast, non-ergonomic activities are completely irrelevant to the ergonomics programme or the functioning of the group (e.g. discussing some production problem that is unrelated to ergonomics).

The attention devoted to group development and maintenance varied markedly between areas A and B and this tells much about their leadership and success as groups. Below we compare the breakdown of the time spent on the five main activities across the four groups in the programme and analyse the leadership styles that led to these group differences.

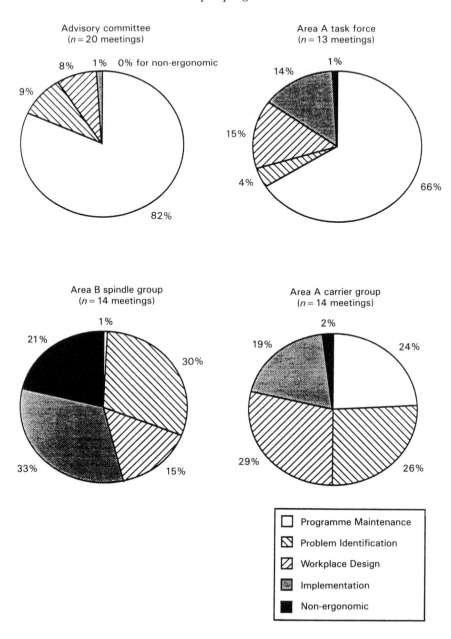

Figure 6.5. The percentage of the time spent on each activity for a sample of chassis plant meetings.

Advisory committee

Overview of group development

For this group the stage of getting their feet on the ground, occurring between March and October 1985, was marked by turmoil. Because this group did not initially take time to discuss their purpose in the programme, they had little focus or sense of purpose. Therefore, they embarked on several agendas that were detrimental to the programme. One, discussed more fully below, was to take it on themselves to analyse some jobs and develop their own ergonomic solutions. This clearly was not their intended role in the programme and they were not successful in this endeavour which was quite demoralizing for the group. However, through strong leadership, a new action plan was developed that served as the framework for the programme over the next 2 years. This framework included a formal statement of mission for the programme and the roles of each of the main groups in the programme, including their own role as advisors and supporters to the task forces.

Figure 6.5 shows the percentage of time the group spent on activities after the up-front development stage. Even after that stage, the group spent a majority of meeting time on programme maintenance activities. As this group's main functions were to guide and steer the programme, such activities are classified here as maintenance, even though the group did not generally focus on its own maintenance (except at two meetings when, after their failure to change jobs, they discussed how to reorient their activities). As the programme matured, new needs were identified to make the programme more successful, for example, involving division engineering in ergonomic designs early in the development of a product.

Leadership

Leadership in this group was strong, perhaps largely because both the plant manager and the industrial relations manager were members of the group. Therefore, when these two sensed a crisis situation approaching, they had the skill and ability to diagnose the problem and to act to correct the situation.

Attendance

The advisory committee had the lowest average percentage attendance (49 per cent). Even so, they devoted very little discussion during meeting time to the attendance issue. Apparently the group did not see low attendance as a problem and did not find it necessary to spend meeting time correcting it. It seems that at this level of the organization, although a high percentage of participants in attendance at meetings were desirable, only a few members were essential to the implementation of their action plan.

Barriers to ergonomics

Except for the brief attempts at job analysis described above, this group was not directly involved in analysis and design changes to jobs. Consequently, the members were insulated from the problems encountered by the other groups in the organization in making job changes.

Task force area A

Overview of group development

Circumstances during this group's early development caused considerable turmoil in its formation and, for a considerable period of time, an unstable existence. This unstable period started when the group began to meet in March 1984 and continued until after June, when a new area manager was appointed. The new manager possessed strong leadership abilities that boosted the programme's importance within the area.

The several-month delay in this group 'getting their feet on the ground' threatened the existence of the programme in the area. However, with the involvement of the new manager and time spent reassessing the area's action plan, the programme took a more positive direction. Activities essential to the programme's existence, like steering the worker group and implementing job changes, were done with more consistency, and members seemed to have a more positive attitude.

The task force spent the majority of its meeting time (66 per cent) on programme maintenance activities dedicated initially to redeveloping their group but later to steering the worker group (see figure 6.5). This was to be expected as the main purpose of the task force was to monitor the worker group for problems and to address those problems.

Because of the requests for job changes from the carrier worker group, the amount of time spent on workplace design and implementation activities was evenly split, 14 per cent and 15 per cent, respectively, and represented a total of 29 per cent of the available meeting time. As the worker group came up with different solutions, the task force analysed them and selected the best solution. Problem identification activities were the responsibility of the worker group, as reflected in the small amount (4 per cent) of time spent by the task force on this activity.

Leadership

Initially, leadership was very weak because of the change in area managers. Middle managers, who were part of the programme, did not want to assume responsibility for the programme until they got approval from the new area manager. Their reluctance continued until the new area manager became part of the ergonomics task force. His strong leadership abilities were very important in the organization of the area ergonomics programme.

Attendance

The average attendance of the task force was 56 per cent. The low rate reflects, in part, the poor start of the programme — average attendance was below 50 per cent prior to the new area manager's arrival, whereas after the new area manager became a member of the group, average attendance increased to 58 per cent.

Barriers to ergonomics

In accordance with the original action plan, an existing job was selected for the analysis of ergonomic risk. Therefore, the task group did not have to wait for the machine sign-offs and quality checks normally encountered during installation of new machinery. In addition, because the line was already in existence, the carrier worker group could go out and look at the jobs as they were running and use the job analysis procedures summarized in the training. However, since most changes had to be made using limited plant resources, specifically maintenance, ergonomic job changes had to wait for other priority maintenance items to be completed. Another constraint that affected the group early on was political in nature. It had to do with whether the group should be responsible for changing time standards after making an ergonomic change. The task force spent several of their early meetings simply discussing this issue which delayed the launch of the area A worker group.

Area A carrier-worker group

Overview of group development

After approving the formation of the worker group by the new area manager, the area task force determined its structure, function, and action plan. Consequently, its up-front development time was short and extremely efficient, confined to two meetings. The main problem to be addressed was training. After the long delay in forming the group after the initial training, it became apparent that members would not be able to remember the necessary information from the training well enough to diagnose problems and to make ergonomic changes to jobs. Therefore, the researcher was assigned the job of retraining the participants.

From the start, this group recognized the need for development and maintenance of an action plan. Therefore, whenever the group recognized a potential problem, they scheduled meeting time to correct it. Afterwards, they presented a summary of their conclusions to the task force for discussion and approval. Figure 6.5 shows that the percentage of time spent on ergonomic activities, including programme maintenance, was fairly evenly split. However, since the programme in the area was designed to divide responsibilities between the task force and the worker group, one might expect that the worker group would have spent very little time on programme maintenance and implementation activities because they are task-force responsibilities. This was not the case. Programme maintenance accounted for 24 per cent of the meeting time, while implementation accounted for 19 per cent. Apparently, the leader of the worker group recognized that for a group to run effectively, time had to be spent on group development and maintenance. Typically, such adjustments were reactions to external and internal factors, including perceived task-force commitment and the speed at which changes were implemented.

Leadership

The quality of leadership for this group was very good. The department supervisor leading the group had an ability to recognize problems before they developed into a crisis and was not afraid to go to the task force for a resolution, all at minimal expense to participation within the worker group. Not only did members believe in what he had to say, but he would also fight for the group's cause, its right to exist, and its right to spend time away from work in order to make changes to jobs.

Attendance

Attendance was very high in this group, an average of 75 per cent. The high attendance rate was due to two factors: good leadership qualities and the belief that the group was changing jobs for the betterment of the worker. This group was at the front line of the ergonomics programme, and the workers in the group were working on improving their own jobs.

Barriers to ergonomics

Since this area chose to work on redesigning existing jobs, implementation of changes, although predominantly a task-force function, had to come from the plant resources under the task force. The carrier-worker group constantly monitored the task force to ensure that its projects were given the priority they felt they deserved. Nonetheless, because of the separation of the task force and worker group the worker group was largely insulated from the problems of implementing changes to jobs.

Area B spindle combined task force/worker group

Overview of group development

Up-front development occurred between February and April 1985. No turmoil or management changes occurred to slow down the process, and the original action plan remained intact, without change, throughout the life of the spindle group. Activities during this time included choosing the work area, training new members to be part of the group, setting up the organizational system, and determining policy.

A major difference between this area and area A was the addition of division support. Since the spindle job was new, installation and refining were supported by considerable resources from the division. Therefore, very little money from the plant had to be spent to make the changes. It appeared that the area manager had carefully planned this strategy to maximize productivity and effectively reduce both the cost and the time necessary to see results. This would enable him to review the results from the programme after a relatively short period of time and determine his commitment to the future expansion of the ergonomics programme.

Figure 6.5 shows the percentage of time spent in each activity after the up-front development phase ended. An extraordinary amount of time was devoted to non-ergonomic activities (21 per cent) and programme maintenance activities (1 per cent) were almost absent. In fact, time that should have been spent on group maintenance was spent on non-ergonomic activities. One possible reason for this behaviour was the number and diversity of the responsibilities of this group. Not only did it function as the task force and worker group in the ergonomics programme, it was also the launch team in charge of getting the machines running. Subsequently, their other launch team agendas often spilled over into the ergonomics meetings.

Leadership

In the light of what has been said concerning the distribution of time in activities, in particular the extremely high amount of non-ergonomic activities carried out during the limited meeting time, the quality of leadership and the leader's commitment to ergonomics has to be questioned. Several times during meetings, the leader should have recognized that the group was not working towards its ergonomics goals and that problems existed. He should have spent meeting time reassessing the situation and bringing the group back on track, i.e. spending time on programme maintenance. Instead, he seemed to welcome slack time as an opportunity to discuss non-ergonomic activities, thereby further aggravating the situation.

Evidence for the quality of leadership was clear during their meeting on 15 October. The entire meeting was devoted to non-ergonomic activities and group satisfaction suffered as a consequence. However, group satisfaction increased considerably after the spindle group met with the advisory committee on 19 October. During the 19 October meeting, the advisory committee reaffirmed their commitment to the programme and asked the spindle group to adjust their action plan to meet several specific programme goals. The setting of concrete goals positively affected group performance which subsequently increased satisfaction.

Attendance

Attendance for this group averaged 59 per cent. This is a higher rate than that of the task force in area A but much lower than that of the carrier group. Since participants in the group had responsibility for new process installations in the area, they were often called away to respond to urgent needs of those projects. Again, this situation reflects the leader's inability to address the problem adequately, and to use meeting time effectively to find a solution.

Barriers to ergonomics

Because the job was new, this group was attempting to make changes during the implementation phase of the production line. The implementation phase presents barriers because of the need to have vendor sign-offs and quality checks before changes are made to machinery. Also, as a launch team, frequent need of participants to respond to other projects severely limited the time available to be spent on this one. Finally, since the job was not yet set up and running the jobs had to be artificially set up in order to analyse them. This was done by the hourly people on the committee who ran the jobs through some simulated trials. Therefore, despite the use of division resources, this group was handicapped in trying to analyse jobs because it had to simulate them and handicapped in attempting to change jobs because it had to wait for sign-off.

Chassis plant ergonomics programme outcomes

Participant satisfaction by group

At the end of each meeting the participants filled out a short form assessing their satisfaction. Table 6.1 summarizes satisfaction with the meeting and the programme in general for each of the four groups in the programme. As expected, participant satisfaction in the area B spindle

Table 6.1. *Participant satisfaction by group for all meetings at the chassis plant†.*

| | Advisory committee‡ (N = 10) | Area A | | Area B |
		Task force‡ (N = 8)	Carrier group‡ (N = 9)	Spindle group‡ (N = 13)
Satisfaction with meeting*	4.3	3.8	4.0	3.7
Satisfaction with programme*	4.3	3.6	3.9	3.6

* Significant between-group difference at the 0.1 level (ANOVA).
† At the end of each meeting the committee members were asked to rate the degree to which they were satisfied with the meeting and the progress of the ergonomics programme. Their answers were based on the following five-point scale: 1, not at all; 2, slightly; 3, considerably; 4, very much; and 5, extremely. Each index is an average of three items.
‡ Ratings by all participants were averaged for that meeting. N = number of meetings rated.

group had the lowest satisfaction scores. The lack of group organization, including assigning tasks, a goal-setting function, may be attributed to the leader of the group. Also, even though events indicated a need to conduct development and maintenance activities, the leader made no attempt to do so.

Area A's task force also had below average satisfaction. As discussed above, the new manager for this group was chosen partly because of his leadership abilities. A comparison between meeting assessment scores before and after the new manager became a member of the group revealed that most scores improved after the new manager appeared. Another characteristic of this group was the relatively high concentration of middle managers. A separate analysis of satisfaction by function showed that the middle managers were the least satisfied with the ergonomics programme (Joseph *et al.*, 1986). To some degree this was simply one more committee assignment on a programme that was not essential to meeting their measured objectives of productivity and quality.

The area A carrier group and the advisory committee generally had the highest scores. As noted earlier, both groups had high-quality leaders from the start which we believe accounted for a large part of their success from the participants' perspectives.

Ergonomic outcomes by group

Table 6.2 presents a summary analysis of the efficiency and effectiveness of the four ergonomics groups in accomplishing their ergonomics objectives. These groups recommended 42 changes to 15 work stations, averaging three recommended changes per work station. Of the 42 changes recommended by the ergonomics groups, 24 (55 per cent) were actually implemented. All implemented job changes were analysed independently by a university researcher using rigorous analysis tools to determine whether risk factors

Table 6.2.   Objective ergonomic outcomes by group at the chassis plant.

	Advisory committee	Area A groups	Area B group	Average/ total
Number of work stations analysed	6	3	5	14
Total number of changes recommended	14	10	18	42
Average number of changes recommended per work station	2.3	3.3	3.5	3
Number of changes implemented	0	9	14	23
Percentage of changes implemented	0	90	78	55
Average time to final decision (days)*	20.5	49	49	38
Average time to implement changes (days)†	—	54	66	60
Percentage of changes using worker participation	0	100	100	43
Practical usability of changes‡	—	2.6	2.5	2.55

* Number of work days between date of analysis of work station initiated and final decision on changes to be made.
† Number of work days between final decision date and installation date.
‡ Ratings of how practical the changes were for operators (e.g. ease of use, and process flow) made by the researcher based on interviews with operators, using a three-point scale from low (1) to high (3) usability.

associated with physical stress were reduced. *In all cases at least one risk factor was substantially reduced* (Joseph, 1986).

Despite this success, we must question why almost half the recommended changes were never implemented. A comparison of the implemented and non-implemented recommendations shows that *all implemented projects involved worker participation*. In contrast, there was no worker participation on 14 of the 18 non-implemented recommendations (78 per cent) — these were all of the 14 projects undertaken by the advisory committee. Cost and engineering feasibility evidence showed no significant difference between recommended changes that did and those that did not involve workers. Thus, these data suggest that worker involvement is an important factor in the successful implementation of ergonomics projects. We caution the reader that the failure of the advisory committee to make changes effectively may be due to factors other than their lack of involvement of production workers. For example, the advisory committee delegated responsibility for implementing workplace changes that they proposed to the ergonomics task forces — groups that already had significant work-loads. That is, the programme was not set up to give the advisory committee authority to delegate work to the task force. However, anecdotal information from plant visits suggested that the advisory committee's failures to implement changes were, at least in part, due to solutions that were technically feasible but not organizationally acceptable to machine operators — a failing that would have been less likely if persons familiar with the job were involved in formulating the solution.

Despite the differences in group functioning between areas A and B, there was no evidence that this led to less effective performance by area B. In fact, area B changed more jobs. However, this difference is due more to the number of jobs available to be changed on the different production lines analysed in the two areas than to the actual performance of each group.

The other main difference between areas was the time it took to implement projects. Area B's projects required resources costing over $10 000 which had to be authorized by division management, taking considerably longer than the less expensive projects of area A for approval and implementation. According to the plant engineering supervisor, implementation of projects costing over $10 000 can take up to 52 weeks, but normally take between 30 and 60 days. Therefore, projects initiated and installed by area B fell within normal plant schedules. Even for projects costing less than $10 000, area B was at a disadvantage as they were working on new jobs for which the vendors had responsibility until sign-off. Thus they had to wait for vendors to make changes.

Table 6.2 also shows the scores for the usability of ergonomic work-station-design changes, by project and group. These ratings (on a scale from one to three) were based on interviews with machine operators after the changes were made. The average score for all implemented projects was 2.5, indicating that changes made through the ergonomics programme were both practical and used by the workers.

Expansion and continuation of the programme

The subsequent history of the programme in both areas was disappointing. Recall that the area A two-tiered committee structure was designed for expansion to other production lines. This would require training supervisors and hourly employees in the new area and getting them started as a 'worker group'. They in fact did do further training; however, the start-up of a new worker group was delayed considerably. Before it got off the ground a new plant manager was brought in who did not wish to continue the ergonomics programme. The area B committee structure was more difficult to transfer. Since a single task force had been established with hourly members dedicated to a particular line, the task force had to be reassembled to incorporate hourly members in a new area. Perhaps for this reason and perhaps because of lack of interest by the area manager, this area did not get even as far as area A in expanding to a new area.

Stages of ergonomic programme development at the stamping plant

The history of the ergonomics programme at the stamping plant is summarized in figure 6.6. Though by many measures the programme can be considered successful, it was not without its challenges. In this section we discuss some of the key challenges and how they were addressed by each of the task forces.

Stage 1 — Laying the groundwork

This early preparation stage took approximately 3 months. During much of this time there was little activity because of difficulties in scheduling meetings that included a large number of very busy people and because of the normal summer shut-down.

With the plant manager firmly behind ergonomics and the prevailing cooperative labour–management climate at the stamping plant, it was easy to 'sell' the notion of a participatory ergonomics programme. At a joint meeting of management and the union, all parties agreed to participate and were enthusiastic about the participative approach. The university researchers made several proposals on the structure of the programme, but it was clear that the area managers had strong ideas of their own with regard to the committee structure and its membership. Because of the similarity of operations within each floor, both area managers agreed that a few representative hourly workers could effectively participate in the analysis and redesign of any operation within their floor. Those hourly workers selected had many years of experience at the stamping plant and had worked on all the major types of operations on their floor. All workers asked to participate agreed enthusiastically.

Through additional discussions the organizational structure designed by the area managers was modified slightly to give the form shown in figure 6.7. The structure consisted of three groups: an ergonomics advisory committee, and an operating committee for each production area (A and B) of the plant. The responsibilities of each group were as follows.

Ergonomics advisory committee: formed to give direction and any necessary project approval to the operating committees. Members of the coordinating committee included the assistant plant manager, area managers, labour-relations manager, industrial-engineering manager, plant-engineering manager and the UAW health-and-safety representative.

Ergonomics operating committees: two operating committees were formed corresponding to the two floors of the plant managed by the two area managers. The responsibilities of the committees were to:
1. identify physical stresses on the operator,
2. brainstorm ways to reduce the stress,

3. select solutions,
4. implement solutions, and
5. implement follow-up.

The groups' members were primarily hourly and salary employees from their respective production areas. They would meet weekly on company time. The most notable difference between the two areas was in the number of production workers assigned to the committee — five were assigned to area A whereas only two were assigned to area B.

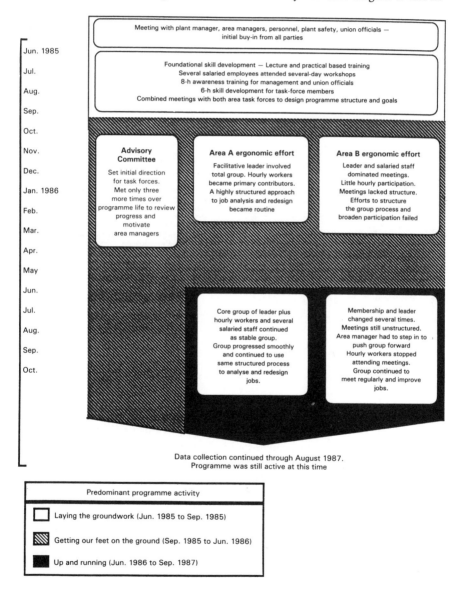

Figure 6.6. Ergonomics programme development at the stamping plant.

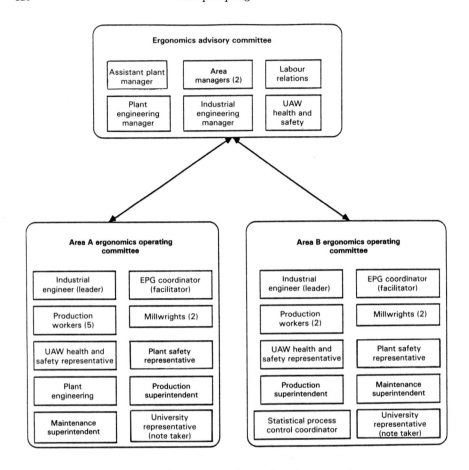

Figure 6.7.   The organization of ergonomics at the stamping plant.

Note that the employee participation group (EPG) coordinators were assigned to the committees. These were specially trained hourly employees assigned full-time to the plant's EPG programme who were responsible for monitoring and facilitating group process.

Some of the salaried committee members, including the industrial engineers leading the committees, had undergone several days of prior training in ergonomics as part of the division-level ergonomics programme. However, for other members 'ergonomics' was a new word. Thus, an in-house 4-h ergonomics training course was held for the operating committees to provide broad exposure to the topic, using a combination of lectures and video tapes of jobs. At this time, most of the participants were happy to participate in the ergonomics programme, but they were unsure of what to expect. They also felt after the training that they still were unable to apply ergonomics. Thus, in a second 2-h training session a simple check-list was presented listing ergonomics risk factors that tend to lead to health problems. Video tapes of jobs were used as case samples so that trainees could try out the check-list. Additional training was provided on an on-going basis by the graduate student assistant assigned quarter-time to work with the task forces.

To begin the process of planning the mission and key roles of the programme, both committees met jointly in a series of meetings. The first meeting resulted in a statement of objectives which included improving work stations to improve health and safety through: identifying problem operations, proposing solutions using problem-solving steps, taking action and following-up. It was decided that productivity would not be a major goal of the programme, though productivity and quality might improve as a by-product. During the second meeting the groups began to add some formal organization. To start the programme off, the EPG coordinators were the facilitators, the industrial engineers were the leaders, and a graduate student from the university assigned to the project was the recorder. However, as the programme progressed the graduate student's role became much more significant as a technical resource to run computer analyses of jobs, provide technical advice on job changes, and provide training in ergonomics.

By the second planning meeting it was evident that the patience of many participants was wearing thin. They wanted to jump in and begin changing jobs. Some of the production workers were heard saying things like 'we should stop all this talking and move toward some action'. There was also a desire for direction from management and some distrust as to whether management would really support the programme. What did management think the goals of the programme should be? Which jobs did they want to see worked on? Would they really provide funding for projects? Thus, the operating committees requested that the advisory committee give them some direction on their goals for the ergonomics programme and the operations which the committees should begin analysing.

The advisory committee met and determined their goals for the programme which were, 'To improve the quality of work life for all operators through work station design'. They decided that the operating committees should begin by analysing new operations that were in the process of installation at the stamping plant. The members of the advisory committee also decided that the operating committees from each area should meet weekly, but they were not required to hold a meeting of set length. The committees should decide their own meeting length week by week based on what they needed to accomplish. It is important to note that, having detected a lack of area-management commitment, the plant manager took an authoritative stance in this meeting and made it clear that each area would devote the resources needed to make the programme work.

Stages 2 and 3 — Programme development and maintenance

In stage 2, the early stage of team formation, the collections of individuals must become functioning groups that develop and learn to use the tools needed to do their work. It is emphasized in the literature on quality circles that early successes are important in this stage to motivate the group to continue (Werther, 1982; Kanter, 1983). For this reason, management support for projects initiated in this phase is particularly important. In stage 3, maintenance of the programme involves maintaining motivation to continue the programme and weathering personnel changes that are inevitable in dynamic, industrial environments. The committees in areas A and B dealt with these challenges in very different ways and with varying degrees of success.

The line between stages 2 and 3 was very difficult to draw in this case. There was no single meeting that represented a turning point wherein the committees finally had their feet on the ground and were shifting to a group-maintenance phase. The somewhat arbitrary boundary shown in figure 6.6 corresponds to a meeting at which the researchers fed back data on group process collected at earlier meetings. Each operating committee used this data as a basis for brainstorming improvements, several of which were subsequently implemented.

Advisory committee

This group met four times over the entire time the programme was studied. After their first meeting when they set the overall direction for the programme, no regularly scheduled meeting times were established. Thus, the committee only met a few more times when they were so requested by the operating committees. Their main purposes at these meetings were to hear reports on accomplishments, provide reinforcement to the committees, and motivate the area managers to continue with the programme. The meetings accomplished these goals well, although in retrospect they were too infrequent.

Operating committee area A

Overview of group development
This group quickly developed a very effective group process which persisted throughout the life of the programme. The five hourly production workers were very vocal and made strong contributions to all discussions. When analysing their first job, several production workers said they needed to try the jobs before they could tell what needed to be changed. Based on this experience they made strong contributions to the analysis and diagnosis of problem areas. The practice of trying out the jobs became a routine part of their analysis procedure (as discussed below).

As the committees began to analyse operations and make recommendations for changes, it became evident that they needed some systematic procedure to follow during this process. The variety of computer tools and enormous quantities of information provided in the ergonomics training were clearly too complex for the groups to use at this stage in the programme and seemed to be best suited for analysis by individuals, not by groups. On the other hand, a purely informal process of looking at jobs, calling out problems, and suggesting solutions seemed to be ineffective. This purely *ad hoc* approach tended to foster domination by a few individuals. It also encouraged skipping over careful problem definition and jumping into problem solutions.

To foster systematic ergonomic analysis by the group as a whole, the facilitator for area B, with input from both groups and University of Michigan ergonomics experts, developed a system which the participants could use to rate the posture, weight and frequency factors which were contributing to the stress (or lack of stress) for each particular operation. Unlike the check-list of stressful postures presented in the training, this was based on a visual representation of a human. The end-product was statistical analysis measurements (SAM), shown in figure 6.8. The raters use a ten-point scale to quantify their *impressions* of the level of stress on particular parts of the body and poor work-station design features where a rating of 1–3 was considered to be acceptable, 4–6 was considered to be questionable, and 7–10 was

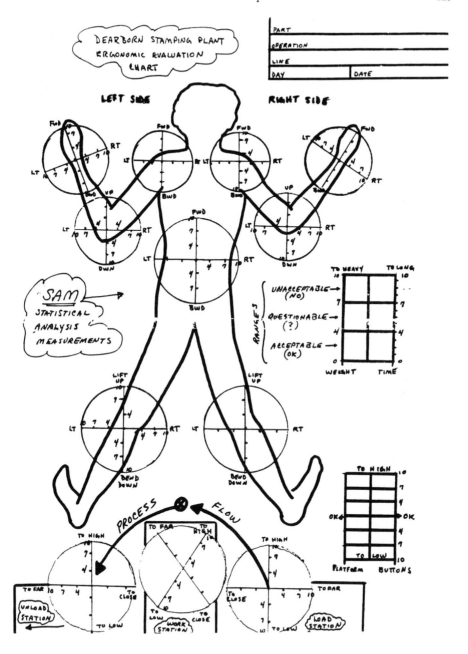

Figure 6.8.   Statistical analysis man (SAM).

considered to be unacceptable. As the committee members entered a numerical value for each partition of the target, they were to consider the worker's posture, the weight of the load, and the frequency of the operation.

Though SAM was developed primarily by the EPG coordinator assigned to area B, it was only the area A committee that embraced SAM as part of a formal job analysis and design process. The process used in area A when they approached a new job evolved into the following.

1. Go to the plant floor where individuals use SAM to rate the operation (hourly members often ran parts on jobs prior to their ratings).
2. Compute and examine the average scores from the participants' ratings (these are summarized in tabular form).
3. Any problem that has an average score of 4 or more is discussed more thoroughly.
4. Invite the operator and supervisor of the job to the meeting to give their input.
5. Brainstorm the problems (not necessarily restricted to those identified by SAM).
6. Brainstorm possible solutions.
7. Determine the best or most feasible solutions(s).

The use of SAM provides a logical rationale for focusing on certain problems and a common ground for individuals to discuss the problem. Each individual has some time on their own to think about the problems prior to group discussion. This is similar to the silent-idea-generation step in the nominal group technique (Van de Ven and Delbecq, 1971) which gives participants time to prepare individually for discussion. This is particularly important to individuals who tend to be quiet in group discussions and increases the probability that they will actively contribute to the discussion.

Only area A invited operators and supervisors outside the ergonomics committee to meetings involving discussions of their jobs. Area A members realized that the operators and supervisors might resist changes to their workplaces unless they were actively involved in the decision process (Coch and French, 1948). As one area A committee member, a machine operator, explained:

> There are people out there, when you say 'ergonomics', watch out. They don't want you out there. Don't touch my job. I know because I was one of them, and I'm on the committee. I didn't want the committee messing with my job. But I did let them and they changed it so it's much better now.

The solution was to invite the operators and supervisors of the operations. The committee analysed to one or more of the weekly meetings to give their input and to understand the committee's reasoning behind the changes they were recommending. Several operators were clearly very comfortable with *their* job as they had been performing it for years and were reluctant to have anything changed prior to meeting with the committee. However, through discussions with the committee and some level of involvement in the changes, they began to understand the reasons for the changes and to 'buy-in' to the solutions agreed upon.

The process described above continued into stage 3 and was still being used by the committee at the time data collection stopped. By this stage the committee still had not become accustomed to using the job analysis computer programs given to them by the university but did recognize when a particular computer analysis was necessary and would request that the analysis be run by the university representative. Another issue in this stage was weathering personnel and shift changes, though the basic core of the group — the leader, EPG facilitator, university representative, and hourly employees — remained stable through the programme's recorded life. Lay-offs, transfers, and new shift assignments made additional training and bringing members up to speed essential parts of the programme. The division-level support of ergonomics courses taught at the university became particularly important at this stage — all new members of the committee were sent to the first course available.

Leadership

The industrial engineer who was selected to lead the area A committee was committed to ergonomics and believed in participative management. Indeed, he was virtually a model participative leader (Maier, 1963; Vroom and Yetton, 1973). The area A industrial engineer expressed his management philosophy, which he indeed acted on, as: 'Total group opinion counts more than my opinion. I believe in the group philosophy. The group operates more effectively than I can alone. These people [hourly production workers] note things that I would miss if I were doing the analysis alone. They know the jobs because they work them every day'.

A high level of rapport between area A's leader and hourly members was established over a long period of time. Even after the programme had been operating for almost 2 years production workers continued playfully to poke fun at the industrial engineer, for example, calling him a 'stop-watch carrier'. By surfacing obvious sources of conflict that have developed historically between industrial engineers and production workers, the traditional social distance was reduced and the potential for conflict lowered.

Barriers to ergonomics

The main barrier was scheduling the time of millwrights to make the changes recommended by the committee. Most of the changes could be made inside the plant with existing resources (e.g. raising platforms, and moving palm buttons). However, the plant was already understaffed and millwright time was scarce. Nonetheless, this barrier was overcome by the commitment of the plant to ergonomics and the motivation of the millwrights and their superintendent who were members of the operating committee. The operating committee painstakingly tracked the progress of projects and often had to push hard to get overtime authorized or get millwright time scheduled.

Operating committee area B

Overview of group development

Developing a group process that fostered broad participation in problem-solving and decision-making was a much greater challenge for the area B committee. From the start, some managers and engineers acted as though the idea of the ergonomics programme was for them individually to find problem operations and determine how they were going to solve the problem. They quickly used their power as technical experts or formal authority as managers (Mintzberg, 1983) to dominate meetings and, as a result other committee members, particularly the production and maintenance employees, took a more passive back-seat role. This continued for a number of meetings and developed into a routine pattern.

The EPG coordinator assigned to the group attempted to break this pattern of domination by a few individuals using commonly recommended techniques for effective meetings (Doyle and Straus, 1976; Jay, 1976; Schindler-Rainman and Lippitt, 1975). He began actively to facilitate meetings, recording ideas on the flip chart and encouraging 'brainstorming' when it was appropriate. Nonetheless, the pattern of unequal participation continued to resurface and maintained itself throughout the data-collection period.

At several points in the programme this committee began to stagnate and experience high absenteeism rates. The area manager had to step in to push the group forward. Eventually, one of the production workers stopped attending meetings completely. The only remaining production worker was a particularly confident and articulate hourly employee who had been upgraded to take responsibility for the statistical process control in his area. The disorganized meeting process continued with a few individuals playing the primary role in problem definition and solution. Nonetheless, because of the dedicated efforts of these individuals, the area B group made many significant changes to operations and was still improving jobs when data collection was stopped.

Leadership

The original industrial engineer who was selected to lead this group was among the best trained in ergonomics in the entire plant. However, he did not have strong skills at facilitating an effective group process. As such, he and a few of his salaried colleagues tended to dominate the meetings. Hourly members were not unhappy, but they were not active and, therefore, their potential contributions were lost to the group. About midway through the data-collection period the industrial engineer was transferred to a different position in the plant and from that time on the leadership was unstable with several different people taking the lead in the group. Ultimately, a more effective leader emerged, a plant engineer, and the group process became much more participative.

Barriers to ergonomics

As in area A, scheduling millwright time to change jobs was a major barrier. However, this group faced an additional barrier not faced by area A. The operations in area B use large dies to stamp out parts and dies are changed depending on the parts being run on a particular day. This meant two things: first, the jobs were not always set up for the committee to observe or for millwrights to work on; and second, a die change might change the ergonomics of the job. Thus, a solution (e.g. raising a conveyor) to a problem posed by a job performed with the die for one part might not be maintained when a different die is set up for a different part.

Summary of area differences in group process

Initially, the two operating committees were equivalent in most respects. They both received the same initial guidance, training, top-management suppport, and available resources, and they both had comparable functions represented on the committee. However, as the groups evolved, a marked difference in the group processes of areas A and B became apparent. Compared with area B, the area A committee made *more effective use of all members of the committee, particularly hourly members*. We cannot simply attribute this difference to a single factor. It may have been the result of some combination of leadership differences, the

Table 6.3. *Observer ratings of the meeting process by area for the stamping plant.*

	Percentage of meetings rated high‡		
Observer ratings†	Area A (*N* = 53)	Area B (*N* = 51)	Significance of difference§
Leadership style			
Discussion was open	77	65	*
Large percentage of people participated in discussion	91	73	*
Decisions made by consensus	98	67	**
Contribution of hourly members†			
Problem definition	88	71	**
Technical solution development	91	68	**
Administrative/financial aspects	40	25	NS
Implementation plan	8	9	NS
Outside operator involvement ‖	Routine	Rare	
Meeting structure			
Percentage of meetings with a written agenda	64	45	*
Clarity of meeting goals	66	43	**
Plan developed for the next meeting	83	41	**
Key people attended meeting	89	75	*
Meeting length (min)	65	45	*

† Most ratings were based on five-point scale: 1, not at all; 2, slightly; 3, considerably; 4, very much; 5, extremely. Exceptions were the percentage of meetings with a 'written agenda' and 'meeting length'.

‡ Percentage high refers to the percentage of meetings which were rated 4 or 5 on the five-point scale.

§ Based on *t*-test of difference between means of areas: NS, not significant; $^*p < 0.05$; $^{**}p < 0.01$.

‖ This refers to the practice of inviting operators whose jobs were being evaluated to meetings. We use a qualitative description as this statistic was not measured quantitatively by meeting.

relatively large proportion of production workers assigned to area A, the structured meeting process used by area A, and the relatively difficult tasks faced by area B who were attempting to change a moving target.

Nonetheless, these group differences were evident from general observations and were also supported by data collected by the researcher present at each meeting. Based on structured observations, the group process for each meeting was characterized by the observer in quantitative form. The results are summarized in table 6.3. Note that the more structured, participative approach to problem-solving undertaken by area A took considerably more time compared with area B. The committees decided for themselves how much time they needed for each weekly meeting. Area A typically went beyond 1 h, while area B finished their business, on average, in 45 min. The question becomes, was the additional time and effort needed for the more participative approach worth it?

Stamping plant ergonomic programme outcomes

Participant satisfaction by group

At the end of each meeting all members of the committee filled out a short form including questions identical to those used in the chassis plant case. Participant satisfaction with each meeting and with the progress of the ergonomics programme is presented by group in table 6.4. Members of the area A committee were more likely to agree that they were satisfied with the meeting and the ergonomics programme generally. It is likely that the committee members from area A were more satisfied with the meetings and felt that the programme was progressing satisfactorily because of their greater involvement in meetings. Thus, one benefit of the leadership approach in area A was more satisfied participants which presumably meant that members were more committed to the ergonomics programme.

Observer ratings of ergonomic outcomes

We have seen that area A used a more formal approach to analysing jobs and developing solutions which tended to involve a larger portion of members of the group in discussion. Their discussions appeared to be more thorough and they also took considerably longer to complete than their area B counterparts. A great deal has been written on how problem solving meetings should be run to achieve an effective meeting (Doyle and Straus, 1976; Jay, 1976; Schindler-Rainman and Lippitt, 1975) and area A clearly conformed more closely to these standards. However, did this 'effective' meeting *process* lead to better ergonomic *outputs*?

One way in which the effectiveness of meetings was measured was to rate subjectively the quantity and quality of progress made on a variety of specific ergonomics tasks in each meeting (e.g. problem identification, developing alternative workplace designs, and selecting the best solution). In addition, summary ratings were made on a five-point scale of the overall quantity and quality of progress on ergonomics projects that each group made in each meeting. The average scores for the overall quantity and quality ratings are presented in table 6.5. Based on these ratings, area A clearly outperformed area B. In a typical meeting area A accomplished more and did a better quality job on identifying stresses, developing alternative solutions, and made better overall progress on ergonomics projects. The differences between the two groups are statistically significant.

Table 6.4. *Participant satisfaction by group for all meetings at the stamping plant*.*

	Rating	
	Area A† (N = 50)	Area B† (N = 41)
Satisfaction with meeting‡	4.1	3.7
Satisfaction with programme‡	4.0	3.5

* At the end of each meeting the committee members were asked to rate the degree to which they were satisfied with the meeting and the progress of the ergonomics programme. Their answers were based on a five-point scale: 1, not at all; 2, slightly; 3, considerably; 4, very much; 5, extremely. Each index is an average of three items.
† Ratings by all participants were averaged for each meeting. These statistics were based on 50 meetings for area A and 41 meetings for area B.
‡ Significant difference at the 0.001 level, *t*-test.

Table 6.5. Observer ratings of meeting ergonomic accomplishments by area at the stamping plant.

	Percentage of meetings rated high*	
	Area A† (N = 50)	Area B† (N = 41)
Quantity of ergonomic accomplishments‡	39	18
Quality of ergonomic accomplishments‡	60	24

* The observer subjectively rated the quantity and quality of ergonomic accomplishments at the end of each meeting on a five-point scale: 1, not at all; 2, slightly; 3, considerably; 4, very much; 5, extremely. Percentage high refers to the percentage of meetings which were rated 4 or 5.
† Based on 50 meetings for Area A and 41 meetings for Area B.
‡ Significant difference at the 0.05 level, *t*-test.

Objective ergonomic outcomes

While observer ratings are one measure of performance, more objective performance measures are also revealed differences between areas A and B. Table 6.6 shows that there was not a large difference in the number of jobs analysed and changed by the committees of area A and area B. As of May 1987, area A had analysed 40 work stations and made changes to 31 distinct jobs, while area B had analysed 33 jobs and changed 27 jobs. However, there was a marked difference in the number of distinct changes made *per* job. Area A made almost three changes per job on average, while area B made half that number. We believe the larger number of changes made by area A is the result of the rigorous process they used to analyse

Table 6.6. Objective ergonomic outcomes by area at the stamping plant.

	Area A committee	Area B committee
Number of work stations analysed	40	33
Number stopped without action	4	4
Number of work stations changed†	31	27
Average number of changes per work station*	2.8	1.5
Average number of work days to final decision*‡	10.4	7.5
Average number of weeks to implementation of changes*§	7.7	3.8
Percentage of sample jobs objectively improved‖	100	100

* Significant difference at the 0.05 level, *t*-test.
† Implemented by 15 May 1987, 23 months into the programme.
‡ Work-days between the date the analysis of the work station initiated and final decision on changes to be made.
§ Weeks between final decision on changes to be made and installation date.
‖ For a sample of five jobs in each area, independent job analysis was performed to measure changes in energy expenditure, posture, two-dimensional static strength analysis, and physical stress as measured using the NIOSH Work Practices Guide. Shown here are the percentages of the ten implemented solutions that substantially improved the jobs by at least one of these measures.

jobs and the involvement of all members. Different members saw different problems in the jobs and brought them to the group's attention. It may be that the area B committee recognized the most serious stresses in their jobs, but some of the more moderate stresses were considered to be worthy of attention by the area A committee.

In both areas project implementation took several times longer than did the detection of problems and development of solutions. The difficulties in getting skilled trades time were the main cause of delays in implementation and a major cause of continued frustration for committees. Area A took longer to come up with solutions and had to wait longer to have changes implemented, in part because they detected and made twice as many changes per job.

Several of the projects (five from each area) were analysed independently by the researcher using four separate ergonomic analysis tools before and after changes were made. The purpose of these analyses was to see if the committees, relying primarily on their judgement, were making changes that were ergonomically sound. By our criterion at least one major type of stress had to be reduced from a potentially harmful level to a substantially safer level, for example, moving from above the 'maximum permissible limit' of the NIOSH Work Practices Guide (NIOSH, 1981) to below this limit, or from above the 'action limit' to the 'acceptable range'. Based on this criterion, did the job changes made by the committee substantially reduce known risk factors? The answer is yes for all of the ten operations analysed.

Summary of stamping plant case

The stamping-plant study demonstrates that the participative approach can work. The programme successfully improved many jobs in less than 2 years and independent job analyses showed that changes were scientifically sound.

Getting the resources needed to change jobs was one of the biggest challenges faced by the ergonomics committees, and on several occasions required top-management intervention. That so many changes were made reflected the management's commitment to ergonomics. As the leader of the area A committee explained, 'ergonomics is a magic word now'. Despite the fact that for years industrial engineers wanted to improve jobs they knew were particularly stressful, it was not until the plant's management made a commitment to 'ergonomics' that the resources for changes were authorized.

The technical resources to analyse jobs using computer models were never effectively developed in the plant. Unfortunately, the plant remained dependent on the assistance of the graduate student assigned to assist them throughout the programme. On the other hand, this assistance was readily available and, at least in the short term, there was no pressure to develop in-plant expertise.

This case also vividly illustrated the importance of an effective group process. Area A was much more effective in utilizing their members, particularly hourly employees, and this resulted in a much more careful analysis of jobs and higher quality work. This case example indicates that, first, the participatory approach is more effective than the more individual-based approach of area B, and, second, creating a task force which includes hourly employees does not automatically make the process participative. Sashkin (1986) emphasizes the importance of effective management of the process of participative management. In the stamping plant case we have seen that effective task-force leadership was an important prerequisite to effective group process.

Summary of chassis plant case

Though the chassis plant committee changed far fewer jobs than the stamping plant, they still made a substantial number of changes that reduced risk factors associated with physical stress. Again we saw evidence that the participative approach can work effectively.

As in the stamping plant, there was considerable variation in the group process across committees. The four groups varied a great deal in how they spent their time, the quality of leadership, and attendance. To some degree these differences were due to the role that the specific group played in the programme. Other differences were the type of job chosen by the group (whether it was existing or new), and the area ergonomics organization (separate task force and worker group or combined). In addition, the types of activities conducted during meetings and the rate of attendance may give an indication of the quality of the group leader and his ability to recognize the need to devote time to group development and maintenance activities.

Throughout the programme, direct assistance by the university was significant. The graduate student assigned to the plant acted as a technical resource and a motivating force to keep the task forces meeting.

Once the organization was developed and in place, maintaining it was essential to its success. Maintenance involved the participants' and leaders' abilities to detect problems or crises and to react to them. Neglecting any crisis could lower participant satisfaction and threaten the existence of the programme. Ultimately, when the initial groups that were dedicated to particular production lines in the plant finished their work on those lines, the plant was unsuccessful in reconstituting the groups to work on other production lines — and the programme ended.

Summary of major challenges, supports and hindrances

Both plants had to deal with a similar set of challenges, and there were both similarities and differences in the supports and hindrances to overcoming these challenges within and between the two plants. These similarities and differences are summarized in table 6.7, divided between 'internal' and 'external' challenges. As for challenges *internal* to the groups, we have seen the importance of effective leadership and group process. The major *external* challenges involve obtaining resources in order to make changes and communicating with outside parties, machine operators, supervisors, management, and the union. These external challenges required on-going selling, training, and follow-up by the task forces to get outside parties for whom ergonomics was not a top priority to provide the needed assistance. Barriers to implementation varied considerably across groups depending on the technology of the production lines they were attempting to improve in the stamping plant and administrative issues of who was responsible for funding the changes in the chassis plant.

Table 6.7. Summary of ergonomics programme challenges, supports, and hindrances.

Key challenges	Supports and hindrances in automotive pilot plants
Stage 1: Laying the groundwork	
1. Top management commitment	1. Supportive, participative management in both plants
2. Initial buy-in from all parties involved	2. UAW and area management buy-in at joint meetings
3. Select committee members	3. Salaried members chosen, hourly workers voluntary
4. Develop foundational skills	4. Awareness lectures, supplemented by hands-on training
5. Designing the mission/structure/roles	5. Area committees met jointly — impatience evident
6. Choose meeting time and place	6. Weekly meetings of about 1 h — stamping meets in shop-floor cafeteria with easy access to jobs
Stage 2: Getting our feet on the ground	
Internal challenges	
1. Clarify member roles	1. Choose leader, facilitator and recorder — leaders key to success
2. Encourage broad participation	2. Managers and technical staff dominated at first. Effective leaders encouraged hourly workers' participation
3. Develop meeting tools and processes	3. Different tools and processes used by different committees. Group process differences correlated with programme outcomes
4. Adjust to shift changes	4. Trained supervisors from both shifts at stamping
5. Group process evaluation and maintenance	5. Survey feedback — salaried attendance on-going problem
External challenges	
1. Demonstrate management commitment	1. Timely implementation = management support
2. Get resources to implement projects	2. Area maintenance on committees and tracked projects. Technology differences in the production lines of different groups influenced implementation difficulties
3. Develop external support system	3. Presentations to advisory committee
4. Involve outside operators and their supervisors	4. Already on chassis committee — one area at stamping invites pertinent operators/supervisors to meetings
5. Communicate ergonomics to the rest of the plant	5. The stamping plant developed buttons, signs and newsletter articles on ergonomics ('word of the future')

Table 6.7. continued.

Key challenges	Supports and hindrances in automotive pilot plants
Stage 3: Up and running	
Internal challenges	
1. Maintain enthusiasm and commitment	1. Project implementation often delayed. Additional training used to maintain enthusiasm
2. Use computer models for job analysis	2. Reliance on university expert throughout programmes
3. Weather personnel/shift changes	3. Additional training to bring new members up to speed
4. Expanding ergonomics to new areas	4. No problem for stamping as committee contained the resources to analyse all jobs. The chassis committees had to change internal membership and failed to do so.
External challenges	
1. Maintenance of external support system	1. Advisory committee pressures area managers for support
2. Implementation of projects	2. Takes constant follow-up by task forces to get resources
3. Management transfers	3. Must continue to sell ergonomics

Conclusions

Should worker participation be used to implement ergonomic changes in the workplace? Overall, the two case studies presented here demonstrate that the use of multifunctional committees, coupled with direct operator participation can be an effective approach. All the implemented projects in the chassis plant involved both worker and management participation. The top-down approach attempted by the advisory committee was *unsuccessful* in implementing change; *none of the changes using this approach were implemented*. We argued that the top-down approach to implementing job changes was ineffective because their solutions often lacked the necessary insights into the jobs that are well known by persons who perform the jobs. Thus, the affected worker was a valuable data resource and his assistance was essential to enhance change efforts.

In the stamping plant, the two ergonomics committees started out similarly but developed very differently. Compared with the more individual-centred task force, which drew less on their hourly-worker members, the more participative task force took longer to meet and solve ergonomics programmes, but by some indicators was more productive. Thus, the benefits of committees do not come automatically. The process must be properly managed.

We have seen that the participative approach can be effective, but is also fraught with challenges. These programmes are not central to getting the product out the door. They are therefore not of foremost importance to the participants, particularly not to middle managers. The provision of training is easy. Getting top management to commit the time and resources necessary on an on-going basis, finding capable leaders, developing and maintaining an effective group process, and keeping participants motivated to continue is a constant struggle.

Clearly, not all groups in the pilot plants were equally successful in meeting these challenges. This begs the question, what steps can be taken to increase the chances of success? We address this question by considering the factors necessary for participative groups to be effective in redesigning work stations and getting their changes implemented. These factors are drawn from more general models of effective problem-solving groups (McGrath, 1984) and take into account the internal dynamics of the problem-solving group, plant-level support, and the availability of resources external to the plant.

Internal task-force factors

Internal group process

In both plants there was evidence that the group process made a difference in participant satisfaction, a finding that is consistent with research on group effectiveness. In groups that more fully utilized members' ideas, participants felt better about meetings and the progress of the ergonomics programme. On the other hand, group research is more equivocal on whether groups with good human relations are more *productive* than groups in which there is less camaraderie (Woodman and Sherwood, 1980; Miller and Monge, 1986). In the chassis plant there was no evidence that the group which paid more attention to group maintenance was also more productive in making carefully designed ergonomic changes. On the other hand, at the stamping plant there was evidence that the area A group, which used a more structured meeting process and involved more members, also did a better quality job on ergonomics. Area A in the stamping case did *not* change many more jobs than area B, but they *did* make twice as many changes to each job which we suspect resulted in higher quality workplaces. This suggests that if the area B committee had been better organized and more participative they would have identified more ergonomics problems in the jobs which they evaluated and would have done a better quality job redesigning these jobs. The key difference between area A and area B, we believe, was not simply a matter of human relations but in the meeting process itself. Hackman and Oldham (1980) referred to this as the 'task performance strategy' of the group. Area A had more effective task performance strategies for evaluating jobs and developing solutions to ergonomics problems. *Thus, ergonomics committees that are designed and trained to be participative and develop an effective group process will keep members motivated and do better quality job redesign.*

Group leadership

Why was area A in the stamping plant so different from area B in their meeting process? We believe that much can be attributed to the differences in leadership of the area A task force in the early stages of the programme initiation. In both plants there were clear differences in leadership across groups. Some leaders were more participative, and sensitive to group process, than others. The question of whether effective leaders are made or born is too complex to be addressed here (Argyris, 1982; Vroom and Yetton, 1973). Suffice it to say that an effective group leader should be either carefully selected or well trained. While technical ergonomics skills can be an asset to a leader, social skills in effectively facilitating group discussion are also central to effective leadership. A second important leadership role that we observed was to act as a liaison to external resources. Leaders communicated progress to management and lobbied for resources that the group needed to accomplish their goals. Thus, *ergonomic task-force leaders should be chosen for their skills at leading the task force in meetings, their commitment to the ergonomics programme, and their political skills for dealing in the broader company environment.*

Group member motivation

Members of the group must be motivated to participate actively. At the chassis plant, middle managers were less enthusiastic about the ergonomics programme than either top management or line workers (Joseph *et al.*, 1986) and at the stamping plant, production workers were more enthusiastic than millwrights. (Participants evaluated each meeting at the end of the meeting. At the stamping plant the skilled trades were, on average, the least satisfied with the meetings and the ergonomics programmes, although this appeared to change over time, particularly in area A where the skilled trades were highly involved and regularly attended weekly meetings over several years.) Production workers are the direct beneficiaries of the changes so it is easy for them to get behind the concept of ergonomics; however, gaining the commitment of middle managers, engineers, and the skilled-trade workers is a more challenging task. Regardless of their jobs, members will be more motivated if they feel that the programme is making a difference. Guzzo (1986) used the term 'potency' to refer to the group's belief that they can be effective. If group members feel impotent, they will quickly lose their motivation to contribute energy to the cause. Their feeling of potency depends heavily on the external support and resources applied to implementing the solutions that they develop. Thus, *to help the task force to feel potent, they should begin by working on simple problems to achieve early successes and be given the external support and resources needed to implement the solutions that they develop.*

Internal plant factors

Top leadership commitment

In both plants commitment by top management at the plant and union leadership was necessary even to begin discussing a participatory ergonomics programme. In both cases a union–management steering committee was set up to provide direction and support to the programme. The exact role of this high-level leadership is less clear (Beer and Huse, 1972). In the cases studied it was neither necessary nor desirable for this committee to get involved in micro-management of the ergonomics programme. In the chassis plant, when the advisory committee took on the task of redesigning jobs they came up with unfeasible solutions and became frustrated with the ergonomics process. But they provided key leadership in defining the overall direction of the programme and ensuring resources for ergonomics changes. The stamping plant managed with very little support from the steering committee who met rarely, but at various junctures called on the steering committee to pressure area managers to support the programme. Thus, *an active steering committee is needed to provide initial direction, motivate middle managers to support the programme, allocate resources for particular projects, and provide reinforcement to the committees doing the work.*

Plant resources for job analysis

The main technical resources needed for analysing and redesigning jobs were video tape equipment, computer hardware and software, and expertise in using these tools. Both plants were very slow in developing these resources internally and relied heavily on university support for formal ergonomic analysis. This would not have been possible if the plants had not been heavily supported by university graduate students (see below), a situation that cannot be reproduced throughout a large corporation. Thus, *advanced training in specific ergonomic analysis tools must be supported by equipment and software acquisition and technical staff must have opportunities and incentives actually to apply these tools at their home plant.*

Plant resources for implementation

The main resources needed to implement recommended job changes in these particular plants were the time of skilled-trade workers and input from workers and supervisors. The large majority of changes were made by millwrights with existing plant resources, hence their time was the critical resource. One reason for putting millwrights and the head of plant maintenance on the ergonomics committees in both plants was to get their support for the changes. If they had not been included, and thus had been asked to make changes to which they were not committed, it seems likely that the project delays would have been far greater than they were. Input from supervisors and workers was also important, not only to get useful ideas, but to gain acceptance for the changes. Thus, *ergonomics programmes should involve the skilled trades, machine operators, and supervisors outside the task force to get them behind the ergonomics programmes.*

Manufacturing-process constraints

The nature of the manufacturing processes faced by the groups influenced their prospects for success. There was a major technology difference between plants which may have contributed to the greater success of the stamping plant programme in changing more jobs. The chassis plant had heterogeneous jobs and, therefore, the production workers on the task force had to be rotated in and out of the task force in order to get their involvement on jobs with which they were familiar, while the stamping plant had homogeneous jobs within areas A and B and, therefore, could assign a few representative workers to a single committee in each area and did not have to reconstitute the committees. There were also important technology differences between areas within plants. For example, in the stamping plant, the press floor was at a clear disadvantage having to change jobs which were themselves regularly changing (as new dies were set up to make different parts), while assembly-floor jobs were captive operations. Thus, *management must be sensitive to the differences in the challenges faced by different plants and different areas of plants and be patient with groups working on more formidable tasks, perhaps even providing extra support to these groups.*

External factors to plants

Corporate support

The division ergonomics programme under which the stamping plant programme was formed would not have happened without the almost evangelical zeal of the top division manager. Support from the division included some funds for implementation and a large training and research grant to the University of Michigan. Through a top-down process, plants were under some pressure to demonstrate they had an active ergonomics programme. The extensive division-level support in body and assembly may help explain the stamping plant ergonomics programme's particularly high level of longevity and success in changing a large portion of the jobs in the plant. By contrast, any ergonomics activity at the chassis plant was internally motivated by visionary plant management who saw the potential benefits of ergonomics. Thus, *in a large corporation, support from division and corporate management can be a major factor in the success of participatory ergonomics programmes.*

External ergonomics expertise

Both plants had exceptional access to university faculty and graduate students. Specially designed training and organizational assistance was provided up-front for the plants, university faculty led a series of meetings to facilitate the design of the programmes, and in each case a graduate student was dedicated one day per week to the plant for several years. The students assisted in technical job analysis, provided on-going training on specific topics, and helped to motivate members to meet regularly. Thus, *external technical support by trained ergonomists on an on-going basis can be an important factor in the success of participatory ergonomics programmes.*

In this chapter we have analysed participatory ergonomics programmes undertaken in two US automotive-parts plants. Comparisons were made within and between plants and there were clear differences in the organization, leadership, and degree of success in making quality ergonomic changes across programmes. Yet the four working task forces in both plants were all successful in making changes to jobs which reduced health-risk factors from an ergonomic standpoint. These cases demonstrate that when plants make a commitment to ergonomics, participatory ergonomics programmes are robust and effective strategies for change.

Acknowledgements

The case studies this chapter is based on were supported by the Motor Vehicles Manufacturing Association and Ford Motor Company. We are grateful for their support of our research, and more importantly, of the application of ergonomics in manufacturing plants. We are also particularly grateful to all the participants in the chassis plant and stamping plant described here who made serious commitments to participatory ergonomics despite their lean staffing and tremendous pressures for production. Both plants also gave generously of their time for the sake of the research process. Tom Armstrong and Don Chaffin were both key actors in securing funding for the study and gave generously of their time and experience in implementing ergonomics in industry. Both were in fact collaborators

in this research and made key intellectual contributions. Finally, we would like to thank
Deborah Liker for painstakingly editing this chapter.

References

Allen, T. J., 1977, *Managing the Flow of Technology*, Cambridge, Massachusetts: MIT Press.

Allison, G., 1971, *Essence of Decision*, Boston: Little, Brown and Co.

Argyris, C., 1982, *Reasoning, Learning and Action: Individual and Organizational*, California: Jossey-Bass.

Armstrong, T. J., Radwin, R. G., Hansen, D. J. and Kennedy, K. W., 1986, Repetitive trauma disorders: job evaluation and design, *Human Factors*, **28**, 325–36.

Beer, M. and Huse, E. F., 1972, A systems approach to organization development, *The Journal of Applied Behavioral Science*, **8**, 79–109.

Coch, L. and French, J., 1948, Overcoming resistance to change, *Human Relations*, **1**, 512–32.

Doyle, M. and Straus, D., 1976, *How to Make Meetings Work*, New York: Wyden Books.

Guzzo, R. A., 1986, Group decision making and group effectiveness in organizations, in Goodman, P. (Ed.) *Designing Effective Work Groups*, Chap. 2, San Francisco: Jossey-Bass.

Hackman, J. R. and Oldham, G. R., 1980, *Work Redesign*, Menlo Park: Addison-Wesley.

Jay, A., 1976, How to run a meeting, *Harvard Business Review*, **Mar.–Apr.**, 43–57.

Joseph, B. S., 1986, 'A participative ergonomics control program in an U.S. automotive plant: evaluation and implications', unpublished Ph.D. dissertation, University of Michigan, Ann Arbor.

Joseph, B. S., Liker, J. K. and Armstrong, T. J., 1986, Group decision making in an ergonomics program in a U.S. automotive plant: correlates of successful meetings, in Hendrick, H. and Brown, Jr, O. (Eds) *Human Factors in Organizational Design and Management*, Vol. 2, Amsterdam: North-Holland.

Kanter, R. M., 1983, *The Change Masters*, New York: Simon and Schuster.

Kelsey, J. L., Pastides, H. and Bisbee, G. E., 1978, *Musculoskeletal Disorders*, New York: Prodist.

Liker, J. K., Joseph, B. S. and Armstrong, T. J., 1984, From ergonomics theory to practice: organizational factors affecting the utilization of ergonomics knowledge, in Hendrick, H. W. and Brown, Jr, O. (Eds) *Human Factors in Organizational Design and Management*, Vol. 1, Amsterdam: North Holland/Elsevier.

Maier, N., 1963, *Problem-Solving Discussions and Conferences*, New York: McGraw Hill.

McGrath, J. E., 1984, *Groups: Interaction and Performance*, New Jersey: Prentice-Hall.

Miller, K. I. and Monge, P. R., 1986, Participation, satisfaction and productivity: a meta-analytic review, *Academy of Management Journal*, **29**, 727–53.

Mintzberg, H., 1983, *Power in and Around Organizations*, New York: Harper and Row.

NIOSH, 1981, *Work Practices Guide for Manual Lifting*, Washington: U.S. Department of Health and Human Services.

Perrow, C., 1983, The organizational context of human factors engineering, *Administrative Science Quarterly*, **28**, 521–41.

Pope, M., 1987, Modification of work organization, *Ergonomics*, **30**, 449–55.

Sashkin, M., 1986, Participative management remains an ethical imperative, *Organizational Dynamics*, **Spring**, 62–75.

Schindler-Rainman, E. and Lippitt, R., 1975, *Taking Your Meetings out of the Doldrums*, La Jolla, California: University Associates.

Van de Ven, A. and Delbecq, A., 1971, Nominal versus interacting group processes for committee decision making effectiveness, *Academy of Management Journal*, **14**, 203–12.

Vroom, V. and Yetton, P., 1973, *Leadership and Decision Making*, Pittsburgh: University of Pittsburgh Press.

Werther, W. B., 1982, Quality circles: key executive issues, *Journal of Contemporary Business*, **11**, 115–24.

Woodman, R. W. and Sherwood, J. J., 1980, The role of team development in organizational effectiveness: a critical review, *Psychological Bulletin*, **88**, 166–86.

Chapter 7
Application of Participatory Ergonomics Through Quality-circle Activities

M. Nagamachi

Structure of participatory ergonomics and ergonomic tools

Participation

The first aspect of participatory ergonomics is participation. Organizing people into a group or team has such advantages as enabling group thinking and cultivating a feeling of involvement. This motivation in the group provides people with the power to carry out problem-solving activities in working environments.

Many researchers have argued for the effectiveness of participation in decision-making related to leadership style. Likert (1967) and Maier (1963) said that participation leads not only to better decision-making, but also motivates people to a higher level of performance. Marrow *et al.* (1967) have shown that impressive increases in productivity can be brought about by providing greater opportunities in decision-making. Hackman and Suttle (1977) have shown that participation in relation to job design is primarily what Herzberg (1966) would call a 'motivator'. The opportunity for participation may increase the involvement of workers in their jobs and may make them more expressive and active in their orientation.

The advantage of participation is dependent on various factors, e.g. the size of the organization and leadership styles. With regard to the size of group, Nagamachi (1983) found that there is a close relationship between group size and group performance in quality-circle activity. He found that groups of six to eight members produced higher productivity in problem solving. Satisfaction and cooperation among members were reported to be highest in groups with five to six members. These findings are shown graphically in figures 7.1 and 7.2. Participation can be maximized by choosing an appropriate group size.

Participation at work can awaken workers' interests in productivity, cost reduction and improvement of product quality. Quality-circle activities in Japan were designed to help cope with product quality. These activities have been implemented successfully in a number of industries.

139

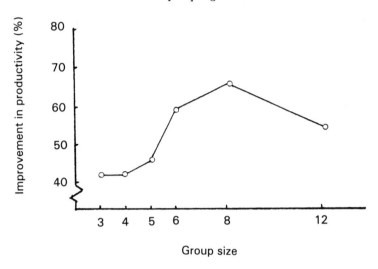

Figure 7.1.   The relationship between group size and improvement in group productivity in quality-circle activity.

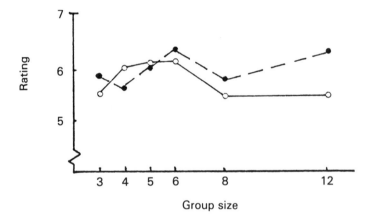

Figure 7.2.   The relationship between group size and satisfaction (○) of and cooperation (●), rated on a seven-point scale between the members of the group.

Quality circles use participation, by organizing small groups of workers into workshops. Since the workers belonging to a quality circle experience and understand the problems of inefficiency, high production cost, and low product quality, they discuss these detrimental factors and express their ideas and ways of eliminating these factors at meetings. When workers achieve higher product quality and higher productivity with their efforts they feel satisfied and experience autonomy and involvement in their jobs. Participation in quality circles not only results in greater production efficiency, but also results in workers' satisfaction (Nagamachi, 1987).

The achievement of both economic gains and humanization of work can be realized if participation in quality circles is combined with job design (Davis and Taylor, 1972). Job-design concepts are similar to those of socio-technical system and work organization. The former emphasizes a harmony between economic and human factors and the latter emphasizes organizational change (Herbst, 1974; Nagamachi, 1985). The concept and procedure of the application of socio-technical systems to participatory ergonomics through quality circles is discussed later in this chapter.

Ergonomic tools

The second aspect of participatory ergonomics is the improvement of the working environment in terms of ergonimic techniques. To illustrate this aspect, this chapter focuses on ergonomic problems associated with the effects of safety, working posture and ageing.

Tools of job redesign for senior workers

The life-expectancy in Japan is rapidly becoming the highest in the world. It is estimated that by the year 2020, 23 per cent of the population will be older than 65 years. Improving the welfare of and providing jobs for older people are current Japanese national projects. According to Nagamachi's (1984) survey of people in their 50s and 60s, almost half were willing to seek jobs for the purpose of increased income or for personal satisfaction. However, ageing affects the physical functions and general ability of older workers, resulting in decreased efficiency and productivity. This is often the cause of management dismissing older workers from their organization. Our concern is to establish a method of job redesign which will enable older workers to continue working in an organization.

The main principles of job redesign for senior workers are:

1. supplement these workers with tools, equipment, or automatic devices which lower the load and stress of the job;
2. implement changes in work processes to make jobs lighter;
3. improve the working environment; and
4. the valuable skills and experience of older workers should be utilized in their jobs.

In accord with the above principles, Nagamachi (1981) established a check-list system (job redesign for life cycle (JDLC)) to identify the factors necessary for job redesign. This consisted of two parts, JDLC-I and JDLC-II. The former part constituted a revision of time study and Turner and Lawrence's (1965) job attributes. Table 7.1 shows a part of JDLC-I. This job-requirement technique has been used in three steel-making establishments to redesign production lines.

The data surveyed in these enterprises were analysed using Hayashi's quantification theory class I (Hayashi, 1976) which is a multivariate analysis for qualitative data. The main factors isolated for job redesign for older workers are shown in Table 7.2: this is JDLC-II, and is intended to be used with workers' participation in job redesign.

Each member of the quality circle checks his or her own job concerning his or her job requirements and draws his or her self-evaluation on a radar-chart (see figure 7.3). After each person has done their job evaluation, all members of the quality circle discuss each evaluation.

All charts are compared in terms of requirements and abilities. The goal of the quality circle is to produce ideas for improving the match between task requirement and older workers' ability.

Tools for working posture

An appropriate working posture is an important factor for workers to be able to do work and to be productive. Frequently, poor working postures in workshops lead to back injury.

Table 7.1. *A job-survey table as used in JDLC-I.*

Name of inspector (................................)   Name of workshop (....)

Notes: enter the frequency of use of each function or evaluation using a five-point scale of load for each task.

Function		Name of task						
		—	—	—	—	—	—	etc.
Finger	Right	—	—	—	—	—	—	—
	Left	—	—	—	—	—	—	—
	Both	—	—	—	—	—	—	—
Hand	Right	—	—	—	—	—	—	—
	Left	—	—	—	—	—	—	—
	Both	—	—	—	—	—	—	—
Arm	Right	—	—	—	—	—	—	—
	Left	—	—	—	—	—	—	—
	Both	—	—	—	—	—	—	—
Legs		—	—	—	—	—	—	—
Working posture		—	—	—	—	—	—	—
Speed of movement		—	—	—	—	—	—	—
Sense	Seeing	—	—	—	—	—	—	—
	Hearing	—	—	—	—	—	—	—
	Balancing	—	—	—	—	—	—	—
Mentality	Judgement	—	—	—	—	—	—	—
	Attention	—	—	—	—	—	—	—
	Stability	—	—	—	—	—	—	—
	Vitality	—	—	—	—	—	—	—
Physical strength	Hand	—	—	—	—	—	—	—
	Arm	—	—	—	—	—	—	—
	Back	—	—	—	—	—	—	—
Work environment	Temperature	—	—	—	—	—	—	—
	Noise	—	—	—	—	—	—	—
	Space	—	—	—	—	—	—	—
	Lighting	—	—	—	—	—	—	—

etc.

Table 7.2. *JDLC-II evaluation chart.*

No.	Job requisite	Definition	Rank Low	Rank High
1	Muscle strength	Degree necessary to carry or push heavy materials		
2	Sense capacity	Visual acuity, hearing, and so forth		
3	Conditions of work environment	Temperature, humidity, high workplace or dangerous level for working		
4	Dexterity	Finger and arm dexterity and speedy activity		
5	Working posture	High-load working posture		
6	Cooperation	Degree of contact with colleagues and cooperation		
7	Attention	Stress and needs of attention		
8	Skills and knowledge	Necessity of skills and knowledge to perform		
9	Experience	Degree of experience to perform productively and safety		
10	Judgement	Specific and general judgement for doing jobs		

(a)   (b)   (c)

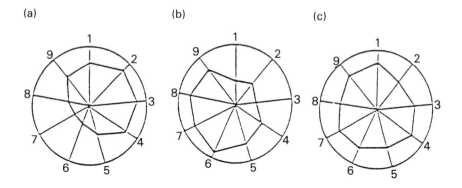

Figure 7.3.   A radar chart for JDLC-II. (a) Job for younger worker. (b) Job for older worker. (c) Job for middle-aged worker.

Nagamachi (1982) extracted ten typical types of working postures from several hundred cases in factories (see figure 7.6): sitting, standing, knee flexion, hip flexion, and variations of these postures. In order to identify the index of the load in each of the ten working postures, Nagamachi recorded electromusculograms (EMGs) of workers, calculated torques through a dynamic model, measured load in terms of relative metabolic rate (RMR) and heart rate, and made a psychological evaluation of load.

EMGs corresponding to the ten typical working postures are shown in figure 7.4. The EMG was measured at the trapezius, deltoid spinal erector, abdominal rectus, femoral rectus

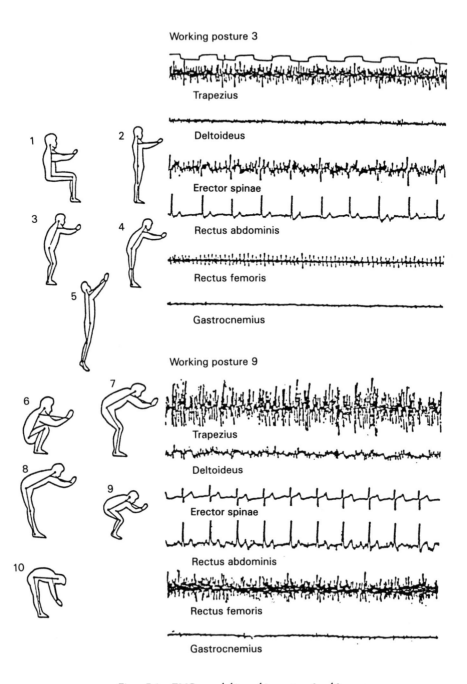

Figure 7.4. EMGs recorded in working postures 3 and 9.

and gastrocnemius muscles whilst the worker was doing a peg-board task in each posture. The EMGs in figure 7.4 show that a strongly flexed posture gives a large burst of activity in the EMG. In particular, in the most flexed posture (posture 9) the highest muscular activity is in the trapezius muscle which keeps the arms raised. Compare this with the lighter working posture (posture 3).

The EMG, heart rate, RMR, torque in a dynamic model and the psychological assessment of work-load were analysed and compared with each other. A high correlation was found among these indices. However, the highest correlation coefficient was obtained between RMR and psychological assessment of work-load (see figure 7.5).

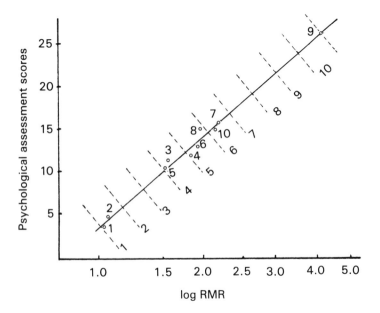

Figure 7.5. Assessment of ten typical working postures in terms of RMR and psychological feeling of load: r = 0.949; p < 0.01.

Figure 7.5 shows a strong linear relationship between RMR and subjective assessments with posture 9 having the highest and posture 1 having the lowest work-load. Accordingly, the space between postures 9 and 1 was divided into ten blocks and each block was assigned a number in a ten-point scale (shown in figure 7.5): each interval indicates a level of work-load. The assessment table shown in figure 7.6 is used as a tool for quality circles to assess the working posture amended according to the aforementioned findings. Table 7.3 shows five ways of improving bad working postures and table 7.4 indicates a procedure for work improvement in terms of working posture. Therefore, figure 7.6 and tables 7.3 and 7.4 are the ergonomic tools relating to working posture, which are applied to industrial situations.

Table 7.3. Five ways of improving working posture.

1. Make the stoop angle as small as possible. The larger the bent angle the higher the work-load

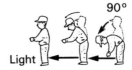

2. Do not force the knees to bend. Bent knees make the work-load high

3. Do not force the body to twist. A bent working posture increases the load, especially if knees are bent

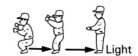

4. Keep the work piece between shoulder and navel height. The height of the arms when peeling an apple is the best height

5. Lessen the distance between the body and the work piece; see comment 4

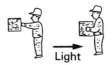

Table 7.4. A procedure for work improvement.

Step 1: Identify the bad working posture
1.   Investigate the working posture
2.   Estimate the work-load of the unnatural working posture

Step 2: Think out the improvement ideas
1.   Investigate the causes of the unnatural working posture
2.   Think out the improvement ideas
3.   Choose the best improvement
4.   Plan the process of redesign

Step 3: Execute redesign planning
1.   Execute the planning
2.   Assess the redesigned jobs
3.   Improve the redesigned jobs if a worse point is found in it

Step 4: Follow up the improvement
1.   Re-investigate the new working posture
2.   Apply the resulting improvements to other lines or plant

New code		Posture number	Assessment score	Definition	Detail
9		9	10	Deeply bent upper body and knees	>90° (irrespective of knee flexion)
8		7	6	Stretching knees and deeply stooping posture	45–90° (back)
7		10		Slightly bent knees and deeply bent back	 0–45° (knees)
6		8		Stretching knees and moderately bent back	 45–90° 
5		6	5	Squat posture	
4		4		Stretching knees and slightly bent back	 30–45° (better than G posture)
3		3	4	Slightly bent knees and back	 0–30° (or standing and slightly bent)
2		5		Stretching posture	A posture for taking something at a high position
1		2	1	Standing Posture	 0–30°
0		1		Sitting posture	 (including creep)

Figure 7.6. Definition of working postures 1 to 10.

Assembly-line improvement by quality-circle activity

The need for assembly-line improvement

As mentioned above, the average age of the Japanese population is increasing; at the same time, however, Japanese workers want to get jobs even in their old age. The Japanese Government, in particular the Ministry of Labour, has begun considering job security for older workers and has enacted the Job Security Act for Senior Workers (1 April, 1986). This law aims:

1. to force enterprises to extend workers' retirement age to 60 years,
2. to secure jobs for workers in the age range 55–65 age within companies, and
3. to support an economic basis for enterprises to follow this policy.

However, the Ministry of Labour wants to know how to secure jobs for older workers. Nagamachi conducted pilot research to determine a method of job redesign for older workers which enables them to continue working. A company which cooperated in carrying out the pilot research is Daikin Industry, Ltd, a well-known maker of air conditioners.

The average age of the workers in Daikin is gradually increasing. Management decided that a company-wide strategy for ageing had to be planned before the average age became so high as to be detrimental. Since ageing will influence a worker's efficiency in any job, all jobs were improved to enable older workers in all kinds of jobs to maintain their productivity. The Sakai plant was selected as a model plant for this pilot research.

The organization of a project team

First, a project team was organized within the company whose task was to ensure the successful execution of the project. The project team was called the Job Redesign Committee for Ageing (JORCA). JORCA was made up of staff from: the personnel, industrial engineering, production management, safety management and welfare departments, the Daikin Labour Union–Sakai Local, the manager and workers of a model line, and a researcher from Hiroshima University (see figure 7.7).

Usually, all members except workers from the line came together to discuss the plan and the procedures. Quality-circle members of the assembly line attended JORCA's meeting when necessary. A model line was an assembly line for middle-size air conditioners; 14 of the 22 work processes in this line were under study.

Training of the project team

The project team received lectures concerning the ergonomics of ageing and techniques of assessing ageing factors. With regard to the ergonomics of ageing, the project team were taught:

1. how ageing decreases human ability and functions,
2. how job redesign helps senior workers by lessening the work-load,
3. how good job redesign enhances productivity, and
4. how the ageing factors associated with high load are identified and how they can be improved.

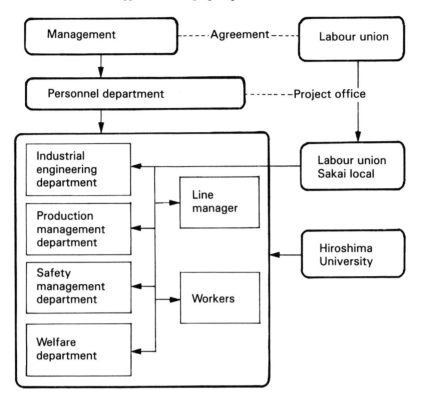

Figure 7.7. The organization of the project team.

Regarding the assessment techniques, members of the project team were trained in how to use JDLC-I (table 7.1) and JDLC-II (table 7.2 and figure 7.3), and how to analyse working postures (figures 7.5 and 7.6). The project team also learned how to set up quality circles to participate in job-redesign projects. In particular, the project team recognized that a radar-chart analysis of their jobs and working posture analysis should be accomplished through quality-circle activity.

The research and analysis procedure

Survey of workers complaints

First, a survey was conducted concerning work-load to obtain information on the type and level of work. The results are shown in figure 7.8. This Pareto diagram shows that 42 per cent of workers expressed a need for an improved working posture; 24 per cent of workers wanted improvement in heavy-parts handling; and 20 per centof the people wanted improvements in walking up and down or long distances.

However, JORCA conducted a survey of about 14 workers' jobs in terms of JDLC-I and identified factors of work-load for older workers. With these findings, JORCA teams

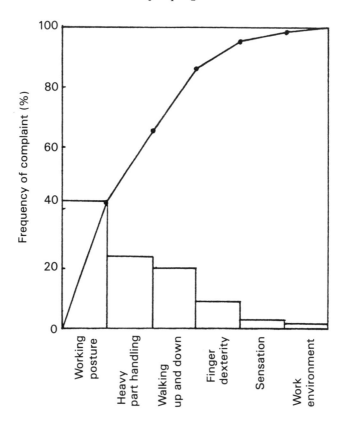

Figure 7.8. Workers' desire for improvements in an assembly line: N = 55, multiple answers.

identified the same problems as the workers. Accordingly, the goals for work improvement decided on were of three main types: working posture, heavy-parts handling, and walking distances.

Job requirement survey

A JDLC-II survey was undertaken by quality circles in order to determine the points necessary for work improvement. The leader of each quality circle explained to the workers how to use JDLC-II and the radar charts. The workers assessed their own jobs using JDLC-II. After the assessment, quality circles compared radar charts and commented on the age-related factors for each member which should be improved.

Figure 7.9 shows the assessments of jobs from process 0 to process 4. The numbers around the circles are the inputs from JDLC-II. Process 0 requires the attention, skills and judgement of a worker, but it also needs muscular strength for handling, which should be reduced. Process 1 requires fast walking and this should be improved also. In processes 3 and 4, the working posture is a problem to be improved. These findings were compared with the results of JDLC-I. JORCA confirmed that there are age-related factors which should be improved.

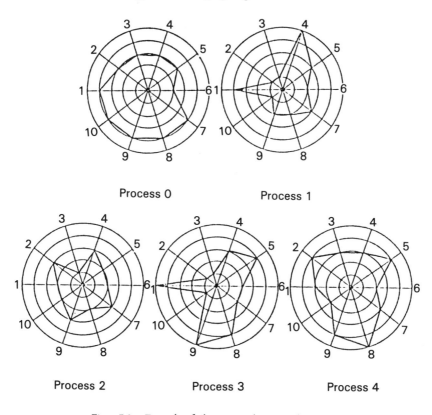

Process 0   Process 1

Process 2   Process 3   Process 4

Figure 7.9. Examples of job assessment by means of JDLC-II.

Analysis of working posture

In the next step of the research, the working postures were analysed. JORCA members took pictures of working posture using a video-tape recorder and classified the types of posture using a personal computer. Workers were given a questionnaire concerning the psychological evaluation of fatigue and painful points on their bodies. The results of the questionnaire corresponded to the analysis of working posture made using video tapes.

Finally, these findings were discussed by the members of the quality circles and JORCA. As a result, 20 improvements to 14 work processes were implemented.

Implemented work improvement

The main work improvements were made in order to give a better working posture, easier handling of heavy parts, and a decrease in the time spent walking up and down. Other improvements were also introduced into the assembly line. The main improvements made are shown in figure 7.10. For example: initially a worker carried a compressor by a hoist at the low position, but after the improvement the compressor is easily carried due to the

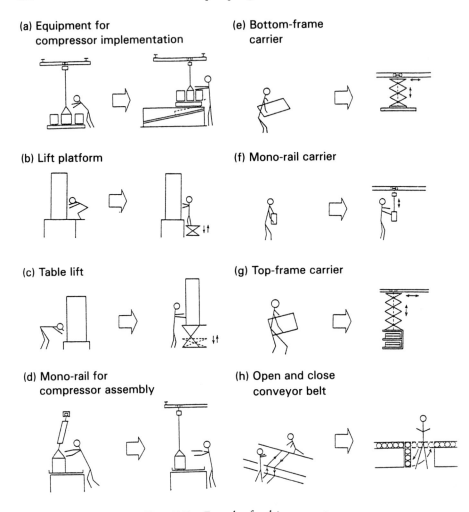

(a) Equipment for
　　compressor implementation

(e) Bottom-frame
　　carrier

(b) Lift platform

(f) Mono-rail carrier

(c) Table lift

(g) Top-frame carrier

(d) Mono-rail for
　　compressor assembly

(h) Open and close
　　conveyor belt

Figure 7.10.　Examples of work improvements.

appropriate table height (figure 7.10(a)); a worker initially had to bend down for a low piece and stretch up for a tall piece, but after the improvement the parts are easily assembled by standing on a platform which moves up and down by means of an oil cylinder moving at $10 \, \mathrm{cms}^{-1}$ (see figure 7.11).

The effectiveness of participatory ergonomics

In order to confirm the results of the work improvement and the effects of participatory ergonomics in terms of quality-circle activity, JORCA asked the workers to execute JDLC-II again and to complete a questionnaire about the effectiveness of the improvements. Of the problems of older workers identified earlier, 11 had been completely removed according to the JDLC-II (see figure 7.12).

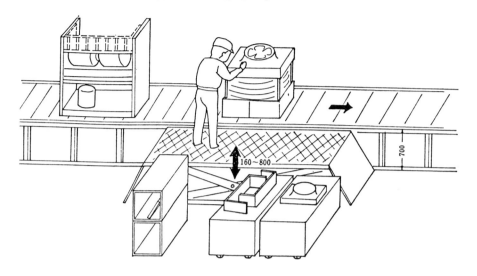

Figure 7.11. A worker working on the lift platform.

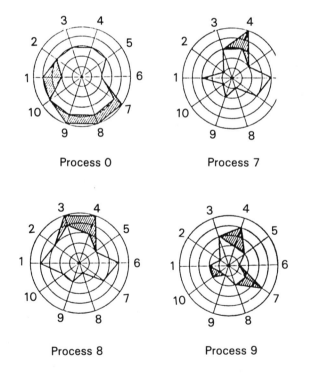

Figure 7.12. Examples of improvements shown on the JDLC-II radar charts. 1, muscle strength; 2, sense capacity; 3, conditions of work environment; 4, dexterity; 5, working posture; 6, cooperation; 7, attention; 8, skills and knowledge; 9, experience; 10, judgement.

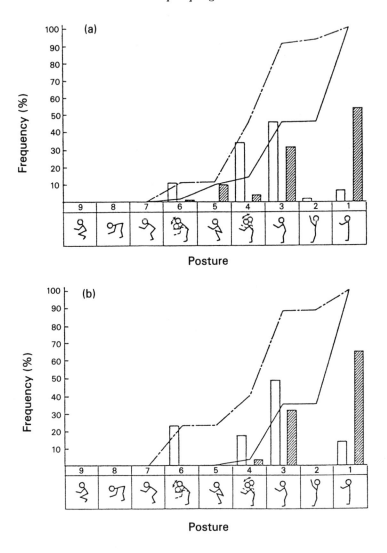

Figure 7.13. Comparison of the frequency of use of a working posture before and after work improvement: (a) process 3; (b) process 4.

Figure 7.13 shows the frequency of use of working postures 1 to 9 for processes 3 and 4 before and after improvement. Improvement resulted in a decrease of high-load working postures in both processes with workers being able to work in the lower level of RMR. A comparison of the average RMR of all processes before and after implementation of the improvements is shown in figure 7.14. Six work processes (0, 3, 4, 6, 8 and 13) were improved a great deal in terms of RMR; the percentage improvement achieved was 16.7 per cent.

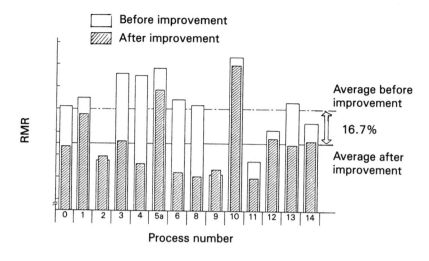

Figure 7.14.   Estimated RMR before and after work improvement.

When asked about their psychological feeling of fatigue after work improvement workers reported a large decrease in work-load and fatigue. Furthermore, they commented that their work was now very light and easy. They also reported that they were satisfied with their level of participation in the project and with their efforts in achieving work improvement.

Finally, a time study executed after improvement found that productivity was 25.2 per cent higher after this participative intervention.

Improvement of working posture in an automobile company

A procedure of group work

The next case of participatory ergonomics through quality-circle activity is the case of Mitsubishi Motors, Kyoto Engine Assembly Plant. The Kyoto Engine Plant assembles all types of passenger-car engine and distributes them to all Mitsubishi group plants. Assembly-line workers in this plant had a problem of back injuries (lumbago) and asked the author to introduce a prevention programme.

First, a project team was organized to carry out the programme. The project team consisted of staff from the health management, industrial engineering and production management departments, an industrial medical doctor and a researcher from Hiroshima University.

The 15 managers of the engine assembly lines also formed a quality-circle. The staff of the health department explained the purpose of the project and gave a schedule of project activity. All members of the project team and managers learned about the ergonomics of working posture, industrial engineering for work improvement, and how to redesign jobs to prevent injury. They were also trained in how to use the aforementioned working posture improvement technique (i.e. figures 7.5 and 7.6 and tables 7.3 and 7.4).

All managers worked together in editing a new manual for working posture improvement and came back to their own lines to teach the leaders of the quality circles with this new text. Quality-circle leaders explained to workers how to identify an unnatural working posture and took the initiative in assessing working postures and in explaining how to improve them at quality-circle meetings.

The manager of each assembly line collected the ideas put forward for better working posture. These ideas were presented in all managers' meetings. The managers and the project team examined each improvement idea one by one from the viewpoints of ergonomic job redesign and the economics of the investment required to improve lines. The method of job redesign and the amount of money allocated was decided for each idea. Implementation was left to each quality circle.

Examples of improvement in working posture

In this section, examples of typical improvements are described.

Adjustable table height

While there are short workers as well as tall workers, table heights are usually fixed. A quality circle in a metal-pattern repair workshop discussed this as a factor in bad working posture. They developed the adjustable work table shown in figure 7.15. The adjustable system is manipulated by inserting a pins into pin holes set at 50-mm intervals. A work table in this system can be adjusted from 550 to 900 mm, compatible with Japanese workers' height. This equipment cost only a very small amount of money since the quality circle made the improvements. Workers reported a decrease in fatigue after introduction of this redesigned table.

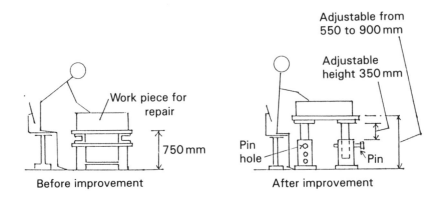

Figure 7.15.   Adjustable-height work table.

Adjustable height of transfer

An improvement was made in a work-piece supply shop in which a worker was required to stack the work pieces on a carrier. This necessitated the worker bending his back in order to put pieces on the lowest layer (see figure 7.16). A quality circle decided to solve this problem by installing an oil lift system. One of the workers found a discarded lift system and repaired it. Therefore, the implementation of the lift system cost very little. This improvement means that work pieces can be easily piled on the carrier by controlling the height of the carrier by simply pushing a button.

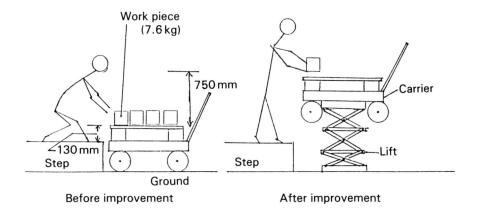

Figure 7.16. Height adjustment of carrier by means of a lift.

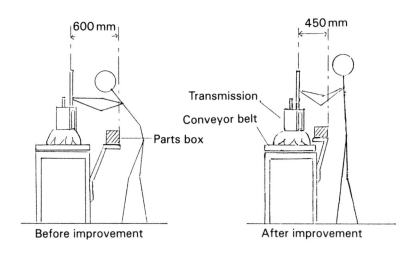

Figure 7.17. Movement of parts box away from the worker.

Improvement of working posture by changing parts-box position

Workers on the transmission assembly line had to bend while tightening bolts to a transmission. One cause of this posture stemmed from the parts box being positioned near to the worker. The workers identified this problem at the quality-circle meeting. Since they recognized that the position of parts box caused the poor posture, they decided to move the box closer to the conveyor belt. The workers had in fact been doing this for many years but had been unaware of this cause until they discussed the improvement in their quality circle.

As shown in figure 7.17, parts boxes were moved 150 mm away from the workers who were then able to stand almost upright. The workers reported a decreased work-load on the back and legs and improved productivity. The industrial engineering line staff confirmed the improvement in efficiency and the decrease in the number of parts dropped. This improved productivity and product quality.

Improvement using a tilted pallet stand

A fork-lift truck carried work pieces for transmission assembly to the side of the conveyor belt where workers picked them up one by one for assembling. However, since the pieces were put on the ground in two layers, the worker had to bend over to pick them up. The quality circle in this line assessed the load of his working posture when he picked up a work piece and took into account the decrease of stress. As a result of the meeting, the improvement shown in figure 7.18 was proposed. The members of the quality circle designed and made the new pallet stand shown in the figure. A tilted pallet for holding work pieces enabled the worker to pick the pieces up very easily. This redesigned equipment enhanced his performance and safety.

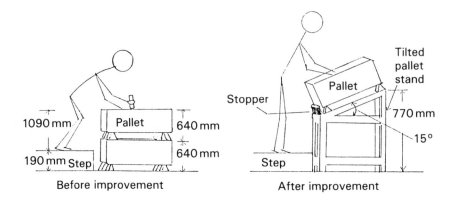

Figure 7.18.   A tilted pallet stand for improving bent working posture.

Changing the position of hand tools

In a cylinder block finish workshop, a worker had to use several hand tools to finish a work piece. Each time he handled a different tool, he had to turn round and twist his body to get the tool. This twisting resulted in low performance. The quality circle discussed this problem and analysed the worker's movement. They suggested moving the hand-tool stand from behind the worker to a position in front of him. A new type of hand-tool shelf was set in front of the worker (see figure 7.19) which made it easier for him to pick up the tools. Furthermore, the shelf is designed to rotate by 90° so that the shelf can be moved when it gets in his way. The shelf enhanced the worker's performance by reducing the stress he experienced due to the twisting movement.

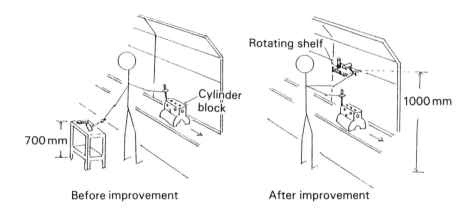

Figure 7.19. A 90° rotating tool shelf to eliminate the twisting body movement.

The results of quality-circle activity

This project for improving working posture was conducted from June to November 1982. During half a year, 224 improvements were completed with the cooperation of all quality circles in the assembly lines (see table 7.5). The most frequent causes of bad working posture were handling heavy parts and poor height of the work table. For handling heavy parts the counter-measures suggested were: redesigning tables, setting the tables at the appropriate height, using adjustable-height tables, using a tilted parts box, and using hoists and cranes. The best solution for the poor height of the work table was to redesign the table. As regards improving working posture the most frequently suggested counter-measures were redesign of table height, use of hoist and crane, and introduction of automation.

Since this project was executed in the Mitsubishi Motor Company, workers have not reported any backache problems. The best problem-solving occurs when those workers who know the problems are involved in the problem-solving process. This is the point of participatory ergonomics.

Table 7.5. A list of improvements to be made to the working posture by a quality circle.

| Cause of bad working posture | Improvement |||||||||||||||| Total | % |
	1 Change position of parts	2 Increasing height of instrument	3 Adjustable table height	4 Redesign of table	5 Foot table	6 Pitting floor	7 Using lifter	8 Using hoist	9 Conveyor belt	10 Lighter wagon	11 Automatic transport	12 Automation	13 Improvement of jig	14 Colouring	15 Change of work process	16 Others		
1 Narrow working area	2			1				8				1				6	18	8.0
2 Inadequate table height		8	8	16	5	3	4	3			1	1				7	56	25.0
3 Handling heavy parts	1	6		12	3	1	2	12	1	13	1	5				26	83	37.1
4 Bad layout of parts	1		3	2			1										7	3.1
5 Bad work position		2		1	6	6	2	1				1			2	10	31	13.8
6 Bad operability of machine		1		1								9	3			2	16	7.1
7 Bad information			2									3		4			9	4.0
8 Bad positioning of switch														1			1	0.4
9 Imbalance of allocation															2	1	3	1.3
Total	4	17	13	33	14	10	9	24	1	13	2	20	3	5	4	52	224	100.0
%	1.8	7.6	5.8	14.7	6.3	4.5	4.0	10.7	0.4	5.8	0.9	8.9	1.3	2.2	1.8	23.2		

Workers assembling a robot — an automation key man system

The aim and procedure of the system

An 'automation key man system' (AKMS) is a new high-technology training system for teaching and training middle-aged workers to make an automatic or unmanned production system. The system has worked efficiently in the Kubota Iron Works Company and the Daikin Industry Company (Nagamachi, 1986). The AKMS used in Kubota is described here in detail. Kubota is a well known maker of cultivators and other agricultural machines. Kubota has succeeded in this system because it was connected to and coordinated by quality-circle activity.

Twenty to thirty skilled workers are selected each year from the production lines and organized into small groups of six members, each acting as a quality circle. In the first 3 months the selected workers learn about the fundamental knowledge and technology related to automation, mechanics and design, automatic control theory, electronics, manufacturing, quality control, and ergonomic safety design. In the following 3 months they practise in the actual manufacturing and electric control process, and then make a real automatic production system to be used in the lines where they work.

After 6 months, the selected workers try actually to make and implement the automatic system in the production lines. After 1 year in training, about half the workers rejoin their lines whilst the others join an improvement team or the industrial engineering department. After the technologically trained workers return to their lines, as foremen or key men, they teach and train their subordinates in the use of the technology and ergonomic safety design. Then, as part of the quality-circle activity, they lead the line workers to make the automation system or to improve the production system which is to be highly automated.

About 300 key men have graduated from this course. These technologically trained quality circles identify inefficiencies and unsafe areas in the present production system, and propose a plan and design for an automated system. To date, these key workers have made many industrial robots, flexible manufacturing systems, and an entire automated production system by themselves. They have made hundreds of pieces of simple mechanized equipment saving many man hours.

These systems and mechanized equipment usually cost very little and move much more accurately because they were made by the workers themselves. Their rate of operation is tremendously high because the workers are very proud of these systems and keep them well maintained, treating them 'like their family'.

Example of a robot and an automation system made by workers

Highly automated system of a crank case manufacturing line

The first case we consider here is an automated crank case manufacturing line where a 20-person line was reduced to a six-person line through automation.

The main problem in this line involved the handling of the crank case — transfer and loading and unloading for manufacturing. An automatic handling system had been completed, but handling and rotating the heavy work piece four times remained a problem. This forced the workers to engage in unsafe behaviour. Two key men and quality circles discussed this problem and planned to make the mechanized equipment shown in figure 7.20.

(a)

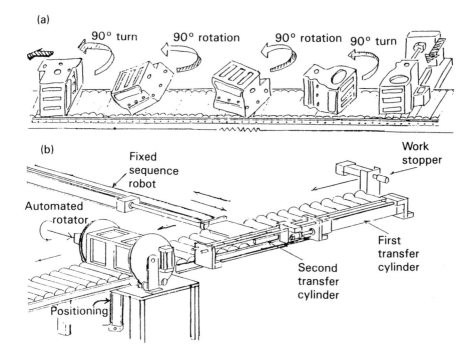

Figure 7.20.   (a) Handling sequence before improvement and (b) automated system after improvement.

The new equipment comprised a cylinder that could rotate through 180° to be implemented in the line. They also completely redesigned all transfer systems in order to improve speed and quality. As a result, a large reduction in the man-power required was realized.

Flexible manufacturing system and robot vehicle

This case comprises a relatively large flexible manufacturing system (FMS) with a robot vehicle transfer system for moving a tractor transmission case which weighs 140 kg within a manufacturing shop. As shown in figure 7.21, a worker had carefully to hang a heavy transmission case on a rotation stand where it rotated and loaded onto a milling machine. After manufacturing, the worker unloaded the case from the milling machine and rotated it again. This procedure had to be repeated for six milling machines. Because of the way the large heavy piece was handled, there was a possibility of accidents occurring and inefficient working.

An automation key man and his quality circle analysed the operation and mechanized the sequence of operations. They then developed a model of the rotator shown in figure 7.21. They made six rotators themselves. After this work, for the purpose of systematizing six milling machines as an FMS, they purchased two robot vehicles to transfer the transmission cases to each milling machine. These vehicles are computer controlled. Consequently, they completed an automated production system which consisted of six milling machines, six rotators and two robot vehicles.

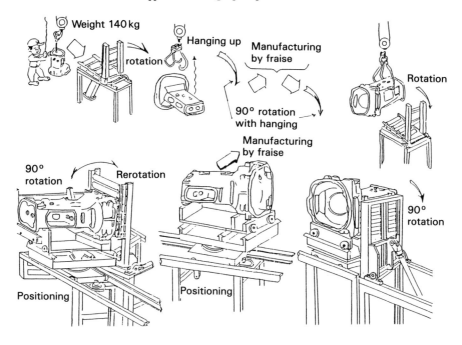

Figure 7.21. (a) Before improvement; (b) after improvement — an automated rotator comprising part of a FMS.

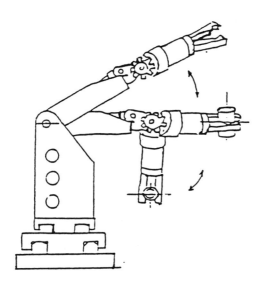

Figure 7.22. 'Hatsuko', a small robot: the drive system is an air cylinder and the positioning refinement is 5/100 mm.

The advantages gained from these activities are a decrease in the possibility of accidents occurring, a decrease in fatigue, and a decrease in the man-hours required to complete the operation. An even bigger advantage is the elimination of a crane operation for handling the heavy work pieces which contributed to the safety problem.

Making a simplified robot and unmanned system

This is a case of a small unmanned production system which was made by an automation key man and his quality circle. First, an unmanned system for assembling parts for a small tractor transmission cover was designed. They made a small robot with four-freedom and a computerized automatic assembly system in which the robot took part in loading and unloading a bush (figure 7.22). This robot was named 'Hatsuko', a girl's name, which means 'the first baby to a young couple'. This experience using a small robot created an atmosphere that encouraged the workers to apply the small robots in many production lines.

As mentioned above, about 300 automation key men were scattered among the production lines and the hundreds of automatic production systems which they made resulted in the great decrease in man-power. This case shows the great power of participatory ergonomics. It is evident that even line workers are able to learn high technology and apply it like engineers.

References

Davis, L. E. and Taylor, J. C. (Eds), 1972, *Design of Jobs*, London: Penguin Books.

Hackman, J. R. and Suttle, J. L., 1977, *Improving Life at Work: Behavioral Science Approaches to Organizational Change*, Santa Monica: Goodyear Publishing Co.

Hayashi, T., 1975, *Theory of Quantification*, Tokyo: Toyokeizaishimposha.

Herbest, P. G., 1974, *Socio-technical Design — Strategies in Multidisciplinary Research*, London: Tavistock Publications.

Herzberg, F., 1966, *Work and the Nature of Man*, Cleveland: World.

Likert, R., 1967, *The Human Organization*, New York: McGraw-Hill.

Mair, N. R. F., 1963, *Problem-Solving Discussions and Conferences: Leadership Methods and Skills*, New York: McGraw-Hill.

Marrow, A. J., Bowers, D. G. and Seashore, S. E., 1967, *Management by Participation*, New York: Harper and Row.

Nagamachi, M., 1981, *Job Redesign for Life Cycle for Older Workers*, Tokyo: Japan Productivity Center Press.

Nagamachi, M., 1982, *A Research Report of Job Redesign, Contract No. 56-2-2*, Tokyo: Ministry of Labour.

Nagamachi, M., 1983, A study of optimum group size of quality circle, *Personnel Research*, **36**, 2–9.

Nagamachi, M., 1984, A philosophy of silver community, *Urban Policy Research Journal*, **4**, 1–33.

Nagamachi, M., 1985, Job redesign for middle-aged and old workers, in Brown, I. D., Goldsmith, R., Coombes, K. and Sinclair, M. A., (Eds) *Ergonomics International 85* pp. 949–51, London: Taylor & Francis.

Nagamachi, M., 1986, *Training and Activation of Senior Workers*, Tokyo: Japan Management Association.

Nagamachi, M., 1987, *Psychology of Quality Circle*, Tokyo: Kaibundo Publishing Co.

Turner, A. N. and Lawrence, P. R., 1965, *Industrial Jobs and the Workers: An Investigation of Responses to Task Attributes*, Boston: Graduate School of Business Administration, Harvard University.

Chapter 8

Participatory Ergonomics — Some Developments and Examples from West Germany

K. J. Zink

Historical Development

Recently, increasing numbers of West German employees from the shop-floor levels of production and administration have become involved in problem-definition and problem-solving processes (Zink, 1990). Problem analysis and solution are natural tasks for executives. For many companies, it is completely new to use knowledge and experience gained directly from workers on the production line for this purpose. This can partially be explained by historical influences. For example, under the Taylorist concept of work organization, a systematic separation between 'muscle and brain' developed. Thought at work was not particularly wanted, and under strong pressure for the division of labour, did not seem necessary. It is not surprising that in Europe the first problem-solving groups at the production-worker level were set up in conjunction with measures to develop new forms of work organization. Economic pressures (in particular lost time and variable costs, but also the high cost of quality control) led in the 1960s in Scandinavia to considerations of how to make production work more attractive by changing the work content. The primary problem-solving approach was to extend and enrich the work content. Supplementary measures involved systematic efforts to improve the work environment (noise, light levels, ventilation, etc.) and the automation of heavily stressed repetitive jobs. These developments were described in West Germany as *Arbeitsstrukturierung* — job redesign. Improving communications and the flow of information within the traditional organizational structure, have led to the introduction of problem-solving groups. These problem-solving groups contain at any one time a group spokesperson, a regularly changing additional member from the working group from either technical or process control, and a manager or foreman.

Publisher's note: this paper was written before the unification of Germany and relates to the area that was formerly West Germany. In this repect, the name West Germany has been retained.

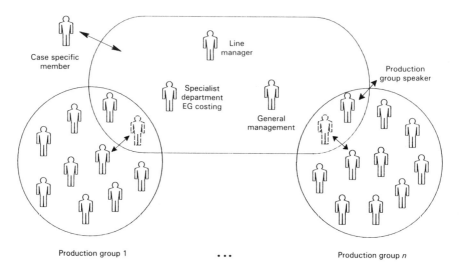

Figure 8.1.   Potential connections between production and problem-solving groups.

Depending on the nature of the problem to be solved, apart from the manager directly responsible, representatives of production control or individual staff departments are co-opted. These groups meet once or twice a month during working hours. The potential connections between production and problem-solving are shown in figure 8.1. The work of these problem-solving groups concentrates on themes drawn from the immediate work surroundings. The themes which are seen as the primary ones for discussion are:

1. questions of work organization (in particular weak points in the production system, but also the problem of filling vacancies created by absence due to holidays, sickness, etc.);
2. questions of environmental stress and organization (e.g. noise, light levels, ventilation and rest areas, etc.);
3. capital investment programmes (acquisition of machines, larger tools, etc.); and
4. the flow of general information (e.g. rationalization plans).

Within the framework of 'work structuring measures' in West Germany, problem-solving groups of this nature have only been introduced to a limited extent. As a result of the 'Japan euphoria' of the last few years, interest in small-group activities has gained considerably in impetus. Research into Japanese management concepts has led to the 'discovery' of the quality (control) circle.

In contrast to the Scandinavian concept, these problem-solving groups are not limited to one or two speakers from the work group. The goal is to involve as many workers from the shop-floor level as possible. Generally, groups of 4–10 volunteer workers would be formed from a workplace under the leadership of a floor section leader (the next responsible management member, for example foreman or *meister*). The groups would periodically meet to identify and examine problems and weaknesses in their own working area with a view to solving them.

The theory behind the introduction of such problem-solving groups is that problems are best identified and solved where they develop and by those who are most involved, i.e. those carrying out the work. For the group to function effectively the group leader's knowledge of his/her role is critical and he/she should attend a thorough course of study before taking up his/her duties. This is a systematic personal development leading to a higher qualification for the individual concerned.

State of the Art

Within the framework of a broad study of the current situation of small-group activities in West Germany, we have researched into how far the work of quality circles includes ergonomic subjects. The empirical data were taken from Ackerman (1989).

Research design

The research was organized in two stages. The first stage was designed to identify how many companies, of more than 800 employees, used small-group activities. A written questionnaire, which was circulated in spring 1985, was targeted at the personnel managers of companies who had shown interest in quality circles through participation in quality-control congresses and seminars. A total of 1201 companies were approached; the response was 60.5 per cent (727 questionnaires). Analysis of the responses showed that:

32 per cent of the companies used small-group activities;
54 per cent of the companies used no small-group activities; and
14 per cent did not respond.

The second stage involved the 235 companies who had indicated involvement with small-group activities and 12 other companies not involved in the first questionnaire. The respondents were asked to comment on the following:

1. feasibility study,
2. initial concept,
3. introduction of plans and strategy,
4. pilot execution phase, and
5. programme development.

The response quota for the second phase which took place in summer 1986 was 65.6 per cent (162 questionnaires). Analysis of the responses showed that:

37 per cent used participation;
12 per cent of the companies did not use participation because:
    such activities were temporarily discontinued (40 per cent),
    such activities had been discontinued indefinitely (17 per cent), and
    such activities were not used or were only used in a small way (43 per cent);
17 per cent did not use participation but did not give reasons; and
24 per cent did not respond.

The nature of the benefits expected by companies from the introduction of small-group activities

The results of the first stage show that the reasons for introducing small-group activities include the improvement of work and the work environment; ergonomic aspects played a subordinate role.

The reasons for introducing small-group activities (in order of importance defined by frequency and weighting) were:

1. improvement of internal communication and cooperation;
2. raising productivity and reducing costs;
3. improvement of the product and its quality;
4. increasing the motivation to work and job satisfaction;
5. better problem-solving;
6. personal development of the worker;
7. strengthening identification;
8. strengthening involvement;
9. utilizing the knowledge and experience of the worker;
10. development of the control and organization concept;
11. increasing the awareness of responsibility;
12. improvement in the organization of the workplace; and
13. promoting the acceptance of changes.

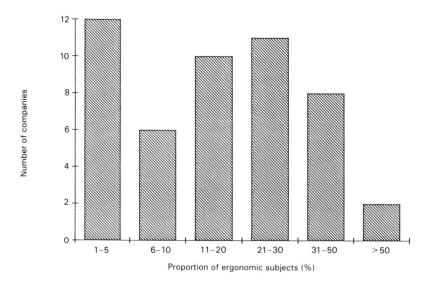

Figure 8.2. Ergonomic subjects as a proportion of the total number of subjects discussed

Subjects discussed by small groups in companies where the concept was introduced without a fixed duration and were subjects where not predirected, were of three types:

1. subjects that involve job satisfaction or dissatisfaction;
2. subjects that involve improvement in communication and cooperation; and
3. subjects that involve an increase in product quality and raising productivity.

Practise shows that small-group concepts which involve an open structure and give individual members a free choice of subject concentrate initially on social themes. Later the group moves on to more complex subjects for which it is easier for the company to provide empirical evidence of economic success.

From the results of the main study it was found that the average proportion of ergonomic subjects was 23 per cent of the total number of subjects. A detailed breakdown of the proportions involved is shown in figure 8.2.

Nature of the subjects discussed

Of the companies questioned: 56 per cent considered human factors/ergonomic themes within the framework of small-group activities; in 12 per cent these considerations were to be introduced into small-group activities; 21 per cent did not consider it necessary to introduce ergonomic themes; and 11 per cent had not examined the subject and, therefore, were unable to express an opinion. Some examples of the themes mentioned by the companies as being discussed by their small groups are:

38 per cent work environment, of which
    37 per cent ventilation and climate,
    22 per cent light levels and type of lighting,
    22 per cent noise, and
    19 per cent miscellaneous;
21 per cent workplace organization
16 per cent equipment organization, of which
    26 per cent lifting and restraining systems,
    21 per cent vacuum cleaning systems, and
    53 per cent miscellaneous;
9 per cent work safety;
5 per cent organization;
3 per cent work process routine; and
8 per cent miscellaneous.

Evaluation of the questionnaire results

In evaluating the results of our questionnaire it is important to bear in mind that the majority of the observed small groups were 'quality circles'. The primary reason for introducing these circles was to define and remove weaknesses in the process and quality of production work. It is important to make a judgement as to whether these quality circles are better described as task forces introduced to deal with a specific problem or whether the group exists as a separate permanent entity. Task forces have subjects predefined by management and have members that, in general, are drawn from various levels within the hierarchy: traditional quality circles have a different membership structure and deal with subjects of different nature. The members are normally drawn from one single working area and from one level in the hierarchy. We have observed that ergonomic themes play a role when the group is allowed to choose its own themes. Practical experience shows that such themes often surface during the initial stages, i.e. bad working conditions are dealt with before other subjects are discussed. Naturally, there are differences between different industries: ergonomic questions arise more often where working conditions are generally worse, for example in the steel industry. The relatively high level of ergonomic themes dealt with by quality circles with free subject discussion does not exclude the existence of task forces at the shop-floor level that have ergonomic themes, in its broadest meaning, as discussion subjects. Such approaches are often to be found during the introduction of new technology or organizational forms. The following case studies show some examples of participatory ergonomics; they show aspects of the quality-circle approach and the task-force concept.

Some case studies in participatory ergonomics

The case studies chosen deal with the following subjects:

1. amelioration of psycho-sociological work stress by integrating shop-floor workers in a project team;
2. an increase in efficiency of the safety delegate through a team concept;
3. the participation element in the introduction of new technology to an office area; and
4. the integration of ergonomic aspects in the problem-solving capability of a quality circle.

Amelioration of psycho-sociological work stress by integrating shop-floor workers in a project team

The information presented in this section was drawn from Slesina *et al.* (1989).

Background

The increasing importance of psycho-sociological work stress makes it necessary to give it corresponding importance in the framework of health and safety at work. In the absence of empirical measurements of health risk it is necessary to develop new methods for defining such risk in the workplace.

One approach is for the affected workers, work health and safety experts and the management to examine the risks to health of the psycho-sociological stress and to consider the possibilities for improvement. The question to be answered is: Which concrete work situations for a particular working group, for example crane drivers, can be linked with particular physical or nervous reactions? The complaints can thereby be considered as an indictor of long-term health risk. The process must be group specific in order that the general connections between work stress and physical or nervous reactions can be filtered out. Through the mixed-team structure (the affected workers, the management, and the expert) a methodology is created for the objective control and assessment of the statements considered by the team. In a second stage the team introduces the relevant improvements. The participation of those affected is considered necessary for the following reasons (Slesina *et al.*, 1989):

1. the definition of the psycho-sociological demands and stress — the experience of those affected providing the primary information;
2. to utilize the knowledge of those affected to develop alternative work practices;
3. to secure the acceptance of the changes; and
4. to register unexpected side-effects of the changes as soon as possible.

Procedure

Experience from two case studies within a steel works led to the construction of two project teams:

1. a melting and pouring team comprising crane drivers and shop-floor workers; and
2. forge-shop teams comprising maintenance and machine operators.

The basis for building the groups was an earlier epidemiological study which showed that crane drivers had a slightly increased rate of chronic circulatory and gastro-intestinal illnesses. In the second group a slight increase in the rate of gastro-intestinal illnesses was registered. A team was created for each shift, giving a total of four teams: crane drivers, shop-floor workers, maintenance, and forge machine operators. Each team met for 12 1-h periods. Work on the problem involved many stages (see figure 8.3) which are analysed below (Slesina *et al.*, 1989).

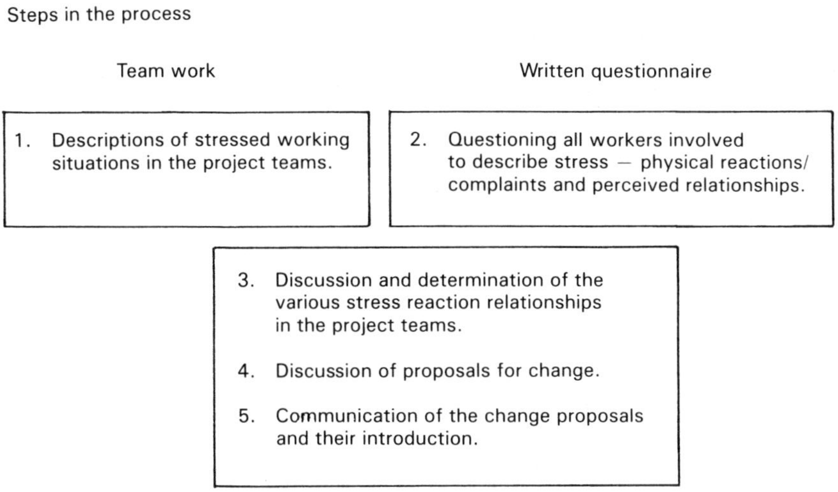

Steps in the process

Team work | Written questionnaire

1. Descriptions of stressed working situations in the project teams.

2. Questioning all workers involved to describe stress — physical reactions/ complaints and perceived relationships.

3. Discussion and determination of the various stress reaction relationships in the project teams.

4. Discussion of proposals for change.

5. Communication of the change proposals and their introduction.

Figure 8.3. Diagram to show the steps used in the process.

Examples of the results and experiences

The results of the work of the group of crane drivers and floor workers (Slesina *et al.*, 1989) are described below.

Step 1

The first step involved mapping the stressful working situations. Each team described between 70 and 110 aspects of the work process that they found stressful. Many of these situations were found to be common to the different teams. This demonstrated that the team work addressed problems of stress that were group specific. The records show that many of the problems described had not been previously identified by the production workers or the health-and-safety experts.

One of the main stress situations described by the crane drivers referred to their responsibility for the security and health of the people working on the shop floor. The responsibility for personnel safety lay with the crane drivers. They took this responsibility under extremely difficult conditions because the workers on the shop floor regularly ignored the traffic regulations. Conflicts between the crane drivers and the floor personnel were therefore a daily occurence.

Step 2

All crane drivers in the company were asked to complete a written questionnaire. The questions were designed to lead to the identification of every working situation in which, for a particular group of workers, an accumulation of physical reactions was correlated.

The crane drivers linked the demands for intense concentration under time-schedule pressure with the following physical reactions: a rush of blood to the head, headaches, increased irritability, internal discontent/nervousness and pains in the neck. Conflict with colleagues and superiors was linked with stomach pains, increased irritability and nervousness.

Step 3

A summary of the results of the questionnaires dealing with the relationship between stress and physical reactions were returned to the teams and discussed in depth. Thereby, actual examples of intense concentration or time and performance pressures were specified. This was achieved by describing concrete work situations where such stress was experienced. In addition, the physical and mental reactions experienced for each situation were categorized.

Step 4

The search for possible changes necessary to achieve a healthier work organization led to 30 to 60 suggestions per team. Some proposals were discussed concurrently by the various teams. Within the crane-driver teams discussion of many of the proposals concentrated on the problem of the intense concentration required to preserve the security of colleagues and the question of accident-free driving in general. The discussions of the proposals included not only the question of people and groups but also technical measures. A complete section of the proposals dealt with the requirements for an improved field of vision for the crane driver through an improved design of the crane cabin as well as through the addition of mirrors and the fitting of closed-circuit-surveillance systems. The main reasoning behind the proposals was that avoiding these conditions of mental stress could lead to a reduction in physical complaints.

Problems involving cooperation with shop-floor workers created many proposals for change. One such proposal was that shop-floor workers should occasionally travel with the crane drivers in order that they understand the difficulties experienced by them. Another suggestion was that formal instruction on communication signals for the floor personnel should be increased; a further suggestion was that, through the installation of communication systems, information exchange between the crane driver and the floor personnel could be improved. Again these proposals involved the avoidance of situations that brought significant stress whilst, overall, the concentration and responsibility required of the crane drivers was accepted.

Step 5

Many of the smaller technical proposals were introduced as the work of the teams progressed; for example, the refitting of the crane drivers' cabins with better seats and better lighting in the central working area. Step by step further proposals were realized. The proposals were made through the company's existing improvement suggestions channel. Many suggestions that require large changes are currently being examined by the management with a view to their introduction.

Increases in efficiency of the security delegate through a team concept

Increases in efficiency brought about via the team concept have been discussed by Hilla (1981).

Background

Safety at work is mainly a management task that involves managers at all levels of the hierarchy. Although the responsibility for work safety lies with the company or with the responsible manager, German law requires the support of safety-at-work specialists and delegates whose tasks are as follows. The safety-at-work specialists (engineers, technicians or foremen) are directly responsible to management for advising the company and its higher management on accident prevention and to advise and support in matters of work safety, in particular through:

1. advice on equipment and furnishings, tools, raw materials and their processing, personal protective devices, workplace design and work procedures;
2. inspection (of equipment and tools);
3. surveillance (through observation, assessing, reporting, proposing, researching, summarizing and evaluating);
4. influence (through information and education); and
5. cooperation with all involved.

Safety delegates must be appointed in all companies with more than 20 employees. The number of delegates appointed depends on the perceived risks inherent in the process, i.e. a chemical process involving pressure or explosive mixtures has a higher requirement for safety delegates than a simple assembly line. Safety delegates are not specialists within the company structure but are in honorary positions and have no special authority in their working area. They are responsible to their immediate line manager and should support him/her by, in particular,

1. convincing and motivating colleagues; and
2. making regular checks on the 'small but important details'.

Practical experience shows that these delegates are often involved in a role conflict. On one hand, without authority, they perform a support function for the company (they are supposed to identify safety risks and to bring them to the notice of the line manager responsible. On the other hand, they are effectively perceived as performing a control function over their line manager and other superiors.

Procedure

In order to confront this problem and to support the position of the safety delegate, a chemical company (the German subsidiary of an international oil company) developed a new concept that, in principle, introduced a broadening of the authority and a furthering of the independence of the safety delegate. The team concept (Hilla, 1981, p. 20) played a central part in developing this new concept.

The safety delegates were organized in groups of workers from specific areas. Eight to ten proved to be the optimum number of people in a group so that absences due to holidays and sickness, etc., allowed for least six to eight members to be available for each meeting. Larger groups proved less efficient because:

1. individuals did not have a full opportunity to express themselves;
2. the subjects discussed remained general because the members came from different areas within the company;
3. the necessary exchange of information with their own line manager failed to take place; or
4. the continuity of the groups could not be guaranteed because the meeting times determined that the groups rarely comprised the same people.

Each group elected a coordinator and his/her deputy for 2 years. At the end of the second year the deputy automatically assumed the role of the coordinator. The election of the coordinator and his deputy must be determined by the group and under no circumstances must it be dictated by the safety expert or the line manager.

Each group should have freedom of choice over the subjects to be discussed from their working area. Potential accidents, insecure safety practices and routines were defined and analysed. The group formed constructive proposals for solving the problems. An immediate reaction on the part of the line managers is highly important.

Each group has the right ask the line manager, the safety expert or other technical experts to attend a meeting in order to obtain particular information or advice.

Each group meets once a month during working time, the meetings lasting some 2 h. Each coordinator drafts minutes of their meeting which are distributed to the other groups, thereby preventing duplication of effort.

Every 3 months the group coordinators and their deputies meet to exchange information with the safety expert; and once a year they meet with the company management.

An example of results and experience

In order to save energy and reduce noise in a refinery, new burners were installed in the oven of a converter. The new burners were larger than the original ones. The following problems were experienced:

1. at one particular point several conduits from the process were laid horizontally under the oven;
2. the space between the conduits and the base of the oven burners was less than 1 m.

The company examined the problem and the suggested solution was to redirect the conduits around the oven at an extremely expensive cost of DM 180 000. The safety delegates decided to examine the problem themselves and suggested an alternative solution costing DM 45 000. Their solution involved ordering the conduits one above the other in the form of a wall, which was better than the company's solution. The team had, therefore, demonstrated the ability to provide a cost-effective solution to a safety problem.

Apart from technical-safety problem-solving, the teams can address organizational and technical–organizational problems. Experience obtained in the initial years of the study showed that, through an increase in individual responsibility, the new concept released an enormous amount of effort and initiative. In addition, area-specific problems were better solved, the contact and information exchange with line managers improved, and the workload of the safety experts was lightened. The motivation of the safety delegates was considerably improved. The frustration experienced by these workers has disappeared.

Participatory elements in the introduction of new technology in an office environment

Background

The increasing use of computers in all areas, not only in business administration, requires an introduction strategy for new technology in order to prevent barriers to acceptance so that the expected benefits of the rationalization can be realized. These barriers to acceptance can have many causes, for example poor 'user-friendliness' of the software or insufficient training. Of extreme importance are the changes in the daily routine created by the use of such equipment. For these reasons it makes sense to convert the affected workers, as soon as possible, into participants.

Process

The example discussed here is that of the introduction of data-processing systems into the administration of the education department of a major company. Five female employees formed the basis of the organization and, as such, were ideal, prior to the main introduction, to work as a participating group. In addition to the five workers affected by the introduction of the new equipment, the group contained the systems manager and a person from the organizations department. The methodology adopted was very similar to a continuous review procedure. In order to document potential changes that would occur during the initial phase and at later stages (for example at the end of technical training), short questionnaires were utilized. This led to an increase in the efficiency of the group work because the questionnaires were used in the sense of a survey and feedback system as a tool in an organizational development process.

The questionnaire used prior to the initial phase contained questions on the following subjects:

1. previous experience with new technology,
2. the expected benefits that the use of new technology in one's own working area would bring,
3. the scale and evaluation of information received to date,
4. the evaluation of the necessity to introduce new technology in one's own working area,
5. reservations and fears,
6. the extent and nature of the participation to date,
7. evaluation of the changes that the introduction of new technology would bring, and
8. evaluation of the motivation potential of the task (7).

At the end of the qualifications training and after an introduction to the system, the questions were asked again and, in addition, particular aspects of the qualifications training were examined.

Practical examples and experience

The results of the questionnaire mirrored the usual experience of such processes. The expectations with regard to the introduction of data-processing equipment were both sceptical and hopeful. The expected easing of the work-load, time saving and better and more up-to-date information must be compared with the fears expressed about the pressure of work involved in the introduction, and the necessity for, but lack of, time available for pre-qualification. Despite a continual flow of information the individuals in the group expressed the feeling 'that they did not really know what was about to happen to them'. This and other statements culled from the first questionnaire led to a session lasting a complete working day where the open questions were systematically dealt with and the group was introduced to the planning of the qualifications work successfully.

The training was carried out by the producer of the equipment, and was supposed to be specific to the work of the group. However, the answers to the questionnaire highlighted the usual weaknesses in such introductions:

1. overall too little time spent,
2. too little time available to become used to the system itself,
3. too little depth in work-specific procedures (too little time spent on working on 'real problems'), and
4. poor-quality instructors.

The broad outline of the results of the questionnaires showed that the systems introduction did not occur without problems. However, through the continuous review procedure it was possible to identify problems in the approach and immediately, together with the group, find solutions to them.

The integration of ergonomic aspects in the problem-solving of quality circles

The following example is drawn from the metal-working industry and shows how a company, coping with problems in quality control, was able to optimize a solution by integrating ergonomic factors.

Background

For the production of fine-cast parts a wax-injection system was utilized. A prerequisite for making these precision parts is the highest possible accuracy in production. The production process creates long process-dictated waiting times. Because of the high production speed, the ejection of wax parts takes place whilst cooling is in progress. The ejected parts fall through a distance of 1.5 m into a collection container (deformation and breakage of the parts are the result). The collection containers are then manually transported to the individual assembly workplaces. During this transportation phase further damage occurs. Following a visual quality check of the parts they are mounted onto wax bars (cluster assembly) (see figure 8.4). Thereafter the procedural organization is characterized by a high division of labour. The work is then split between two workplaces:

1. machine operation, control and de-burring; and
2. assembly tasks.

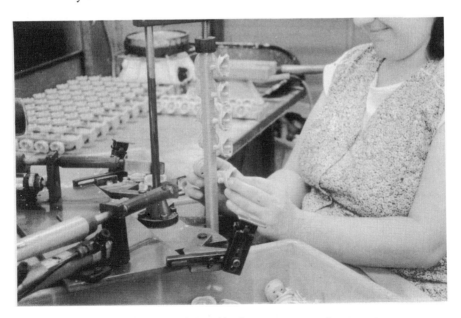

Figure 8.4. The assembly of wax parts onto a wax bar.

Process

Problem definition
The quality circle operating in this area decided to examine the subject of 'the assembly of wax parts from injection machines'. The basis for the problem-solving process was to define the relevant production times and, thereby, calculate the production cost through the agreed production plan. In addition a map of the work machines was drawn.

Target definition
From the problem definition the following potentially worthwhile targets were identified:

1. a reduction in wastage,
2. an increase in quality,
3. a reduction in the process-specific waiting times, and, thereby,
4. a reduction in the production cost.

Causal research and evaluation
Using the usual problem-solving techniques of quality circles the causes of problems were quickly identified. Because the opportunities for improvement in the production process were limited, the group concentrated on weaknesses in the production organization.

After examining a number of approaches the following solution was agreed upon (in the same way as the problem was chosen):

1. the cluster assembly should take place directly at the wax injection machine; and
2. the wax parts should first fall into a catching arm and the necessary pneumatic connection should be organized such that the part is passed to the collecting container at slow speed after which the arm returns at high speed, thereby preventing heaping in the collection container and manual transport to the assembly workplaces.

This solution was effected by the company's own construction department.

Execution
First, a practical experiment in cluster assembly at the wax-injection machine took place with a particularly breakable and pressure-sensitive part. It was necessary to install a small work table with assembly tools and a glue gun. Furthermore, a styrofoam box for transporting the assembled clusters had to be designed. The intention was that the clusters should be transported by means of a conveyor belt running directly past the workplace.

It was immediately obvious that the available workplace was impractical for the cluster assembly because:

1. there was no place to store complete clusters or the necessary tools;
2. the space was too limited; and
3. there was interference in the sitting assembly process through oil mist from the adjacent machinery.

Therefore, the following changes were made:

1. the working area was increased by removing one machine from the production line and reorganizing the remaining machines;

2. relaying the electrical wiring and compressed air lines;
3. the working table was reconstructed taking into account anthropometric and ergonomic constraints (foot room/reaching plan), the construction plan for this being created by the group;
4. the working tools were constructed (the container necessary to transport completed clusters was reconstructed in order that a problem-free transport on the conveyor belt could be guaranteed);
5. the working plan was reorganized; and
6. protection devices were constructed to reduce the stress on the sitting worker created by the oil mist.

In the subsequent group meetings with, additionally, a representative from the company construction department, the proposal was re-examined. The reworked proposal was then introduced. The new production times and production costs were recalculated and the results evaluated by the group.

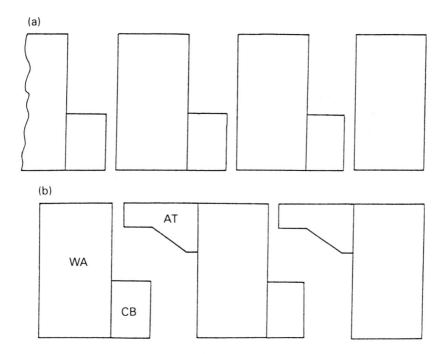

Figure 8.5. Machine floor plans: (a) prior to reorganization; (b) after reorganization. WA, wax-injection machine; CB, control board; AT, assembly table.

Examples of the results and experience

This example shows that members of quality circles without specific qualifications are in a position to solve ergonomic problems that occur in their own working environment.

However, it must be said that specialist departments must give technical support where necessary.

Control information showed that, for each newly constructed workplace, a monthly saving of approximately DM 1400 was possible. Thus it can be seen that economic constraints and ergonomic constraints are not mutually exclusive. The integration of ergonomic considerations initiated by the group leads naturally to a higher acceptance of the chosen solution and, thereby, to an increased satisfaction of the workers' own working conditions.

Evaluations and conclusions

The use of worker participation, whether general (organization of work content) or specific (questions of ergonomics) is not yet widespread in West Germany. This is because worker cooperation and influence are only occasionally found and then as isolated examples. This is partially due to the managerial perception that the participation of workers in the organization of their own workplaces involves increased costs. One of the case studies presented here shows that this is not necessarily so. Problems created by the introduction of new technology, or forms of organization, can, by including those workers affected in the preparation, be solved and considerable amounts of money can be saved. Participation can reduce barriers to acceptance and thus increase effective utilization.

Other themes, such as improvements in work safety, that with traditional procedures show poor results, can be handled very successfully by means of worker participation. It must be stated that the participative approach requires a suitable environment. An effective participative leadership concept is necessary (Zink *et al.*, 1989). Where this is present it is simpler to solve ergonomic problems, in other cases the burden of proof must lie with the cost effectiveness of the solution chosen.

Acknowledgement

The author would like to thank Mr N. J. Vickers, Kaiserslautern, for the translation of this article.

References

Ackerman, M. P., 1989, Quality circles in the Federal Republic of West Germany; impeding and supporting factors of a successful realization, in *Europäische Hochschulschriften: Reihe 5, Volks- und Betriebswirtschaft, Bd 1033*, Frankfurt: Peter Lang AG.

Hilla, M., 1981, Increased effectiveness of safety delegates through new concepts, *Sicherheitsingenieur*, 11, 18—20.

Slesina, W., von Ferber, C., 1989, The integrated stress-strain concept — a challenge to sociology to contribute toward health-promoting work design, *Zeitschrift für Arbeitswissenschaft*, 1, 16–22.

Zink, K. J., 1990, Quality circles — still a topic?, *Personalführung*, 3, 147–153.

Zink, K. J. and Ritter, A., 1989, Improvement of the quality of working life through problem solving groups (Quality Circles, Lernstatt-Gruppen), in *Werkstattberichte Humanisierung des Arbeitslebens*, Bonn: Bundesministerium für Forschung und Technologie (BMFT).

Chapter 9

Linking Perspectives: Using Macroergonomics to Make Technology Work in Organizations

M. M. Robertson and S. M. Dray

Introduction

It has become well-known in both management and technical circles that organizational factors play a key role in the effective use of information technology. Recently, this issue has been highlighted in management journals, such as the *Harvard Business Review*, and in seminars, in addition to being a topic of discussion among managers and users of computers. When organizational factors are minimized, technology is often less than successful. For instance, Bowen (1986) has reported that American businesses have spent billions of dollars on office automation, yet white-collar productivity is no higher than it was in the late 1960s. There still remains a persistent and troubling gap between the inherent value of office systems and the ability of organizations to use them effectively (Leonard-Barton and Kraus, 1985). This gap is due, at least in part, to inadequate attention to the organizational change which always accompanies technical change. Often management must change work roles and communication patterns, in addition to installing new office technology, to obtain the strong positive impact which technology can have on business. Knowing how to get high productivity from new office technology is a learning process that takes time. Many companies are still attempting to learn the process. However, the present authors and others (e.g. Mumford, 1983; Pava, 1983; Bowen, 1986; Freedman, 1986; Hoerr *et al.*, 1986; Sullivan, 1986; Ciborra, 1987; Robertson and Dray, 1988) have found that to be most effective:

1. the major impetus for technology introduction must come from management; and
2. technology must support real business goals.

An on-going dilemma in the corporate world is how to design technical systems which truly support these business goals. It can be very difficult for managers and technical experts to communicate these goals accurately to one another, and even more difficult for them to speculate on the organizational impacts of their decisions. Without this clarity, it is difficult to plan and proactively manage new and existing systems.

To integrate the perspectives of managers and technical experts, an increasing number of organizations are designing or using methodologies based on the principles of socio-technical systems (STS) (Emery and Trist, 1960; Mumford, 1983; Pava, 1983) or macroergonomics (Dray, 1985; Hendrick, 1986). The conceptual underpinnings of these two approaches are basically the same. Systems must be designed jointly to optimize both technical and organizational needs. The goal is to maximize the acceptance and effective use of technology within organizations and to minimize its potential negative impacts (Dray, 1985). To accomplish this, some companies have used participatory strategies including participatory design, user involvement, and participatory ergonomics. Each of these strategies is based on the end-user being involved in decisions regarding those aspects of technology implementation and use which directly affect them, including participating in decisions about workplace design, job design, computer systems design or design of other organizational system characteristics.

User involvement in the design and implementation of new information technologies is critical. Workers in the area of 'humanistic' organizational systems development (e.g. Argyris, 1957; Likert, 1961; McGregor, 1960; Lawler *et al.*, 1981; Mumford, 1983; Lawler, 1986) have discussed why user involvement or employee participation is important. Lawler (1986) suggested that there are three reasons why employee participation is important in today's world.

1. Individuals are becoming more specialized in their work activities, acting more as problem-solvers who make pertinent decisions. At the same time, new technologies allow people to have greater control over their work. This means that 'effective performance' is not always measurable by traditional standards since it is more difficult to measure the quality of a decision than to measure the number of keystrokes per hour.
2. Changes in the work-force are occurring as people are generally better educated, more knowledgeable about workers' rights and hesitant about accepting 'orders' simply because of a person's position of power.
3. Many people want control over their work. Lawler emphasizes that participation moves information, knowledge, rewards, or power down the organizational line, which allows people to have more control over their work.

Similarly, Mumford (1987), in a project to design a new expert system, found that members of an expert system user design group developed a sense of control over collective decisions and events as well as a better understanding of how the organizational environment could be controlled. Levels of enthusiasm, motivation, and excitement increased within the user group as participants got closer to attaining the objective; group members also felt more competent and confident of their skills and abilities to build an effective expert system successfully. Therefore, participative systems seem to foster positive user attitudes which, in turn, enhance implementation of the new technology (Lawler, 1986; Mumford, 1987).

One methodology based on these macroergonomic principles was developed at the Honeywell Corporation. The Human-Technology Impacts Department was initially asked to explain the impact of technology on corporate culture for the Chief Executive Officer. This department was involved in modifying a traditional systems development process that would incorporate proactive planning for human and organizational factors of new technology. The result was a methodology called DELTA (delivering education to leverage technology application) (Dray, 1987). DELTA is based on the assumption that, to be fully successful, technology implementation requires integration of three separate perspectives: business, technical and organizational.

The DELTA process and guide consists of three components. First, cross-functional collaboration between three major groups (user management, information services, and human resources) is of major importance. Second, there is a linking of the organizational and 'people' tasks that need to be managed with the traditional project phases using a flexible framework. Third, a set of tools helps to accomplish these organizational tasks. This specific set of 'tools' can be tailored to technologically stimulated organizational change. A major benefit of the DELTA process (mindset) and guide (which contains the framework and tools) has been the increased cooperation between professional communities which previously were unaccustomed to working together.

This chapter describes the development of DELTA, discusses the way in which this new methodology was designed, and provides examples of its use within a variety of Honeywell divisions.

The Honeywell Experience

In 1984, Honeywell's Corporate Executive Officer expressed concern about the potential impacts of sophisticated technology on Honeywell's business activities and the corporate culture. The vice presidents of Corporate Information Management and of Corporate Human Resources supported an internal study to assess the long-range impact of information technologies on Honeywell's corporate culture and interaction with customers and other business associates. A project team led by Corporate Information Management and supported by Corporate Human Resources was formed.

The project team, chaired by the second author, began by interviewing 160 people including representatives of all levels of Honeywell (that is, 'end-users' to top executives). In addition, outside the company, this team interviewed people in academic life, government, as well as other organizations. Project team members collected other data by reviewing business and research publications, commissioning several independent studies and conducting internal focus groups. Furthermore, they surveyed users in an effort to ascertain the impacts of Honeywell's own internal voice messaging system on communication patterns. In all, five summary observations were recorded.

Summary observations

Management is not powerless in the face of technology

The values of top management must act as a catalyst to initiate and guide the process of technology introduction. The active role of management is essential if an organization is to attain optimum results. As one executive stated, 'You need a leader, a manager who pulls the process along instead of waiting to be pushed'.

Technology does not 'stand alone'

Technology must be integrated into organizations, and its very introduction creates change. Managing technology, therefore, means managing change. New information technology can alter the climate, chemistry, and character of the workplace (Mainiero and DeMichiell, 1986). The smallest changes occur in the hardware system itself when computer technology is introdueed into the workplace. Potentially far more significant and widespread is the impact of such technology on the organization.

For instance, using computers as tools of personal productivity or for communication can precipitate changes in the ways people think and interact (Kelleher, 1986). There is now a radical change in traditional work practices. The old 'scientific management' method of dividing work into discrete tasks that require little skill or training is becoming obsolete for reasons relating both to changes in the work-force as well as changes in technology. In a computerized workplace, many functions can be completely integrated by the computer (Hoerr *et al.*, 1986). We now realize that it requires cross-functional teamwork to manage technology in order to perform the new automated tasks. Thus, it is pivotal that management accept the responsibility of integrating office systems into the organization (Freedman, 1986; Sullivan, 1986).

Technology impacts organizations as a magnifier

Problems tend to get worse, while effective organizations tend to get better. Linda Edsel, director of systems implementation for Holiday Inns, states that 'If an organization has a communications problem, or if it is understaffed, or if employees are generally poorly trained, a computer system will only magnify the problems caused by those conditions' (Zemke, 1987).

The human element is as crucial in the introduction of technology as the technical element but is easier to ignore in planning

Ignoring the human element in planning is disastrous since a system which people cannot or will not use will fail as surely as one which is technically deficient. New technologies cannot fulfill the promise of increased productivity and reduced operating costs if employees are not able to use them and adapt to resultant changes in the office environment. Too frequently, management decides to adopt new technology resources without adequately considering and planning for their human resources (Mainiero and DeMichiell, 1986). The National Conference on People, Organizations and Office Technology reported '75% of the companies surveyed are experiencing significant problems because of insufficient attention to "people problems" in the introduction and use of new office technologies' (Westin *et al.*, 1985). When organizations fail to consider the impact that new technologies may have on employees, they can find that the benefits of increased productivity and lowered operating costs are slow in coming (Taylor, 1984; Mainiero and DeMichiell, 1986).

There is no 'one right way' to introduce technology

Organizations need flexible solutions. Each organization defines its business goals and organizational structures uniquely, and the method of implementing new technology can vary widely according to specific situations. Hence, there is a necessity for a 'generic' or flexible methodology that allows companies to work through their problems systematically and arrive at the optimal answer for their specific needs.

The Honeywell project team recommended that, in addition to raising the awareness of the previous observations, a specific set of 'tools' and a flexible methodology which could be tailored to specified situations were needed. The team was given the mandate to develop these in the second phase of the DELTA project.

A second task team was formed to develop a new methodology for introducing information technology — one which proactively addresses the human and organizational impacts of the technology throughout the project life cycle. This task team included experts from human resources, information services, and human factors. Given adequate time for team development, this cross-functional team worked extremely well together.

The management of any such team is crucial. Since there are few shared norms, words, and assumptions, for cross-functional teams, more time must be spent in helping team members to understand and appreciate both the differences and similarities in their perspectives. This additional time required for the team to coalesce is typical for mixed teams (Bowman, 1986). Because effective cross-functional teamwork is so crucial to effective technology implementation, we discuss here (1) the relevant expertise required, (2) the barriers, and (3) making collaboration work. We then describe the DELTA methodology in depth.

Relevant expertise required for cross-functional teams

Early in the project, the team recognized that the task of technology introduction is itself a cross-functional task. Individuals on the team had seen from experience that technology introduction was most likely to succeed when three perspectives were considered from the outset: business, technical, and organizational. It was therefore imperative that the knowledge of these functional groups be integrated. The team proposed that one way to do this was to form a cross-functional team with members from these different perspectives to manage the entire process. At different times in the project, specialists from other disciplines or resource areas could be brought in to assist in the project.

The type of expertise required within an organization varies with each project. Individuals may be available and have the knowledge and expertise, but they may not have the corresponding functional responsibility or authority. In some organizations, having this functional authority and/or responsibility may not be necessary in order to provide resources for the project team but in other cases, authority may be necessary along with relevant expertise.

Business perspective

Technical systems must have a clear and well-communicated business reason for being implemented, or they are likely not to succeed. A distinct link to the business strategy is crucial, and managers must understand that they play a major role in demonstrating that link for their organizational components of system success (e.g. politics, personnel/staff characteristics and skills, preferred management style, and informal/formal norms). These types of organizational characteristics need to be recognized and utilized proactively in system design to smooth the process of acceptance.

Technical perspective

Technical expertise is required in the actual technical design of the system (hardware and software) as well as in the earlier phases of developing technical specifications, prototyping, and later in testing. Development of both hardware and software (e.g. user interface) design are two particularly critical skills which may be unfamiliar to some more traditional technical experts. Prototyping is also a way to develop and test new systems which requires a somewhat different set of skills, but which greatly eases the integration of the technical system into the organization.

Organizational perspective

'Organizational experts' who are needed to ensure that the system can and will function effectively within organizational constraints are often knowledgeable in organizational development, organizational communication, organization/human needs assessment, organizational design, job design, staffing, compensation, labour relations, training and education, user involvement, performance measurement and appraisal, career development, and strategies for the adaptation and adjustment to change. All these areas are potentially critical to successful organizational adaptation to new technology.

Barriers

Bowman (1986) and Dray (1987) have pointed out that merely having the inputs from these three perspectives does not ensure a successful project. The project team itself must work together to understand and leverage each others' expertise. There is usually a high degree of interdependency between the different functional groups which cannot be ignored. Effective collaboration across these three groups, however, is not always an easy task. Each may have unique and specific goals, objectives, priorities, schedules, politics, language, values, norms, and reward structure. Specifically, the technical systems experts are often primarily focused on technology and its design; the human resources professionals usually direct their attention to the people of the company; while the line managers are typically concerned with the business objectives and goals and the 'bottom line'. Also, the reward structures within each functional group are often based on individual and functional team performance, not on the performance efforts of participating on a cross-functional collaborative team. All these issues contribute to the difficulties of a cross-collaboration effort.

   A second barrier to cross-collaboration (Bowman, 1986; Mumford, 1987) is the failure to recognize the potential value and contribution of forming a cross-functional team. This is often due to a lack of understanding of what each discipline and department can contribute to the successful implementation of technical systems. The expertise of individuals within each group may vary widely, and their potential contribution can be significant. However, unless this is communicated to the other groups, this knowledge will be overlooked.

   Another inhibitor to collaboration is the set of stereotypes that functional groups tend to have about each other based on previous history and preconceptions. For instance, Bowman (1986) reported these types of comments: 'Those organization development people just come in and stir things up!'; or 'The techies are so enamored with the technology that they ignore the poor end-users!'; and 'Nobody bothers to really understand my business needs'. On the other hand, prior knowledge of other groups also can facilitate the cross-collaborative process. If, for instance, a manager has had previous experience with a helpful professional from either human resources or information services, the value of that professional's participation may be more obvious to the manager, and vice versa.

An additional factor in communicating cross-functionally is understanding the working language within each discipline. The jargon and terminology that each area has are unique and may be unfamiliar to other members of the project team. Often, a term used by one group may have a completely different connotation for another. An example is the term 'processing'. To technical systems experts, 'processing' means the technical processing of data. For organizational, development experts, 'processing' typically refers to dealing psychologically with an event, idea or feeling. 'User involvement' is another term that is commonly used and often misunderstood; for the technical systems expert, 'user involvement' means having management of the user organization involved in the task, whereas, the human resources expert envisions all potential end-users providing input and comments. Moreover, the line manager may think it means utilizing a representative team of users.

Decision-making and thinking styles among the technical and human resource groups also are quite different because of their inherent value systems and the individual differences of the employees. For instance, a technical expert is usually very familiar and comfortable with highly structured, detail-oriented tasks. The systems development methodologies allow the expert to analyse their technical and systems problems systematically. In stark contrast, the human resources experts are less structured, more dynamic and process-oriented.

In addition, individuals typically select one of these professions based on what type of style is most comfortable for them. There is some evidence (Dray and Monod, 1986; Schelling, 1980) that these two functional groups do typically think differently and prefer different organizational environments, which may explain why it can be difficult to develop a cross-collaboration team.

Making collaboration work

It is clear from the above why collaboration between different functional groups does not occur naturally or easily, and why it must often be orchestrated. Many authors (Bowman, 1986; Ciborra, 1987; Mumford, 1987) have stated that there are important elements for establishing and creating a project team. Some of the more important elements are described below.

Commitment

The commitment of top management is important. Time for the collaborative effort must be allocated and assumed to be part of each project member's job responsibilities and performance standards. There should be adequate rewards for being involved in the project team. Having a clearly stated business strategy and vision are important if the team is to continue to be motivated. Top management's continuing support and interest often necessitates frequent contact with the project, especially when business strategies and priorities are changing rapidly. This ensures that the project will continue to evolve to meet changing business needs if the business environment itself is in flux.

Selection

With the support structure developed, specific individuals need to be selected to serve on the project management team, and it is usually their responsibility to establish and set the tone of the collaboration process. Experience with prior cross-collaborative efforts is helpful, especially if it has been with the specific functional areas involved in the project. Furthermore, the team should ensure that there are some commonalities among the team members (e.g. analytical skills, intellectual ability, and value of teamwork), as this allows for communication to be established more readily.

Organization

The project management team organization will vary according to the organizational situation and the personalities involved. The work unit manager may be the leader or it might rotate periodically. Each project management team member might have their own project teams to handle specific tasks, or there could be a single project team with which the functional specialists consult (Bowman, 1986).

Team facilitation

Since difficulties can easily arise with this type of collaboration, a facilitator may be helpful or even necessary during the initial stages of the project management team development. This individual generally needs to have an acute awareness for potential differences among the group members and experience making synergistic use of these differences. The facilitator can monitor the potential miscommunication and conflict that may arise within the team. The facilitator can also assist the team in developing shared goals, clarifying roles, and establishing norms; all of which lead to a more effective team. It is possible that the facilitator role can be taken by one of the three project management teams; but this may prove to be difficult since it can be difficult to focus on the group process while trying to be involved in the work content. If an outside facilitator is selected, this individual can concentrate on the group process of the project team, especially in the initial start-up phase. They can also monitor the team's progress more easily.

Team development

It takes time and experience working together to develop a well-functioning, effective and efficient team. This is true for all teams, but it may take significantly longer for cross-functional teams. System- and task-oriented groups sometimes have more difficulties in collaborating, since they are frequently unaware of the time period that it takes to develop and establish rapport with other group members and learn about each others' knowledge and skills. Building trust and respect takes additional time to develop among the team members, but once achieved, efficiency and productivity rises sharply (Bowman, 1986; Mumford, 1987).

The structure of the DELTA framework

The DELTA framework was developed to link all three perspectives (business, technical and organizational). The team divided the framework into nine project phases, intended roughly to approximate those used by the technical community:

Phase 1: identify problem/opportunity,
Phase 2: identify project resources,
Phase 3: define user requirements,
Phase 4: select hardware/software,
Phase 5: acquire/develop new system,
Phase 6: pilot the system,
Phase 7: implement throughout organization,
Phase 8: evaluate the system, and
Phase 9: maintain/support the system.

Within each of these project phases, there is a parallel list of human and organizational tasks called 'organizational management actions', which are linked in time and in order to steps in the technical cycle. These are actions which experience and research have shown to be important in effectively managing organizational change (Kling and Scacchi, 1980; Mumford, 1983; Pava, 1983; Ciborra, 1987). They include actions by line managers and by human resource professionals, often in concert with technical experts. Organizational management actions cover the following key areas:

1. *Link to business strategy.* Make sure it is clear and well understood throughout the organization.
2. *User involvement.* Ensure participation of appropriate users in decisions (technical and organizational) which impact them.
3. *Key-player involvement.* Ensure participation of so-called 'key players', inside and outside the organization. People whose buy-in is important because of positive power, or political factors.
4. *Employee communication.* Make sure there is a high degree of communication.
5. *Organizational assessment.* Find out what the current organization is like and identify potential changes and their impacts.
6. *Organizational design.* Consider structural aspects and potential changes.
7. *Job design.* Identify potential individual and collective changes in jobs.
8. *Ergonomics.* Make sure the workplace 'fits' the physical characteristics of the employees.
9. *Training and education.* Make sure employees know how actually to use the technology and understand how, conceptually, the technology fits into the business.
10. *Personnel processes.* Make sure changes in staffing needs, pay programmes, benefit packages, etc., are considered and implemented if necessary.
11. *Individual transition management.* Make sure individual employees have adequate support during the process of transition to the new system.

In addition, there is a list of tools which are helpful in accomplishing each organizational management action. Each list is a 'smorgasbord' of available tools which can be used in conjunction with each other or by themselves.

The remainder of the DELTA guide is devoted to extensive documentation. This is consistent with the Honeywell culture. Both the organizational management actions and the tools are described in detail in order to give the user an idea of why particular actions need to be considered, who needs to be involved, who should use each of the tools, the benefits and disadvantages of each tool, and where the tools can be obtained. Some examples of the phases, the organizational management actions, the tools and back-up material are given below.

One example of a typical organizational management action for phase 1 is to conduct an informal assessment of the organization. The purpose of doing this is:

1. to take the organization's strategies, climate and values into consideration before committing to a change effort involving the technology;
2. to identify which organizational elements should be preserved and supported;
3. to identify potential barriers to be overcome or avoided;
4. to determine how receptive the organization is to the change; and
5. to help define the project and its scope, so that it has the best potential for success (i.e. appropriate estimates of project length, required resources, etc.).

Some of the steps that might be taken to accomplish this include:

1. conduct an informal organizational assessment by reviewing existing data and by informally interviewing top and middle managers;
2. identify such things as organizational culture and values, the strategic direction of the business, the strategic direction of the information services, the strategic direction of human resources, and change readiness.
3. if unionized personnel will potentially be impacted, involve your labour relations specialists in the assessment. Depending on circumstances, personnel may need to participate in interviewing, or may identify self union participants;
4. be particularly alert to shifts in business strategy, climate, and changes in management throughout the project life cycle, since these are critical elements in project support, funding and acceptance; and
5. do a more formal assessment of these same organizational factors after the project has been authorized and form a project team if this seems necessary.

The people responsible for these tasks are usually the managers with support from the organizational development specialists. Often the 'champion' of the project, the person who is primarily involved in 'selling' the idea to the organization, will do the assessments in order to identify organizational supports and barriers prior to formal authorization.

The DELTA guide itself is intended to be just that, a guide, but not a 'bible' or a methodology to be applied rigidly. It is meant to be used by a team representing the three perspectives. Although it is not absolutely necessary to have such a team explicitly formed in all projects, we have found that it is crucial to have access to key representatives of these perspectives during the planning of the technological change to make sure that critical elements are addressed.

How this approach has worked

Pilot sites

The DELTA process and guide were completed in summer 1986 and pilot tested in projects from two different settings:

1. large cross-functional order entry system in one of Honeywell's large innovation divisions; and
2. a local-area network (LAN) in a small department within Corporate Information Management.

Honeywell's experiences in these initial locations corroborated the importance of cross-functional input, but their experiences also suggested that the initial assumption (that the most effective way to achieve input was by use of a cross-functional team) was incorrect, at least at Honeywell. In both cases, the team approach was not as effective as hoped, and the project team found other ways to ensure this type of input. For instance, in the order entry system case, the project manager was an extremely talented and proactive leader, well aware of the organizational impacts of technology. She was able to draw upon her own knowledge and very real political skills to assure that input was given where needed. She found the guide to be useful in managing political aspects and validating political processes of the system's design and implementation.

On the other hand, the LAN project was smaller, and there were inadequate resources in the human resource organization. In addition, it became clear that the crucial need was to help the department identify how the technology could help them meet their business purpose. Since this was a technical department, it was particularly important for them not to lose sight of this in favour of a 'fun' new technology. They also had a proactive manager who refined the departmental 'vision' of their business purpose and sought appropriate input. In both cases, the leadership was crucial to the eventual success of the system.

Further experience

Since the pilot tests, Honeywell has used the guide in a number of ways. An initial team training was held during the pilot phase. Three divisional teams were trained to work as cross-functional teams and to use the guide itself. Although the training was well-received, these teams have reconfigured and have used the materials in different ways than the project team initially expected. The Information Services Department in one division incorporated the organizational steps into their own systems-development methodology. The second division cancelled the project soon after training. However, the team members still use parts of the training materials and guide in other projects. In the third division, the project has been stalled, but the team has continued to function more informally. This has confirmed the project team's suspicion that, at Honeywell, specific team formation is not effective.

In another innovation division, the organizational development staff was trained to use the guide. In turn, they have used it to implement a factory management system (MRP-II) on a short schedule. They also used the guide to put together a human resources plan to accompany the other technical project plans. In addition they used members of the

Human–Technology Impacts Department as consultants to help them raise awareness of these issues within the user groups.

In divisions with no organizational development expertise, the Human–Technology Impacts Department has provided consulting to specific projects or helped the division to locate specialists in the area. In addition, the department has worked with information service organizations in identifying organizational implications of different projects. The projects have varied in size, scope, and type of application. Although each case has required DELTA to be modified, the basic approach has served as a good guide.

What we have learned

The DELTA process and guide were developed by a cross-functional team to address the issue of how to manage the organizational impacts of new information technology proactively. They have been tested and used internally to aid a wide variety of technology implementations. While it may not be necessary to have a formal team, there needs to be input from three different perspectives (business, technical and organizational).

It is interesting to speculate on why the cross-functional teams did not work as effectively at Honeywell as had been originally anticipated. The cross-functional teams themselves identified a number of reasons, including the difficulty of working together, time constraints, and lack of rewards for working in this new way. The organizational culture needs to support cross-functional teamwork; there should be adequate rewards to reinforce this cooperative relationship. Indeed, we have noticed the impact of cultural norms of cross-functional teams in different companies. These types of teams appear to work much more effectively in a company which makes teamwork an explicit value.

Making technology work within organizations will require increasing methods of linking together diverse perspectives and bodies of knowledge. The specific methods for implementing these technologies will depend on many variables — the most important probably being organizational culture. Those cultures where teamwork is actively supported will probably find it easier to use explicit cross-functional teams. Those cultures where is is less true will need to find other mechanisms, including more explicit methodologies or more formal ways of getting the input from multiple groups. Whatever the mechanism, organizations will need to confront and deal with the organizational side of technological change proactively. By doing so, they can realize the full benefits of technology which have proved so elusive. Macroergonomics can be a powerful aid to such companies, but, in the end, it is management's quality and commitment to the organizational integration of technology that will determine its success.

References

Argyris, C., 1957, *Personality and Organization*, New York: Harper and Row.

Bowen, W., 1986, The puny payoff from office computers, *Fortune*, **26 May**, 20–4.

Bowman, B. L., 1986, Cross-functional collaboration: teaming for technological change, in Brown, O. and Hendrick, H. W., (Eds), *Human Factors in Organizational Design and Management II*, pp. 511–15, Amsterdam: Elsevier.

Ciborra, C. U., 1987, Reframing the role of computers in organizations — the transaction costs approach, *Technology and People*, **3**, 17–37.

Dray, S. M., 1985, Macroergonomics in organizations: an introduction, in Brown, I. D., Goldsmith, R., Coombes, K. and Sinclair, M. A., (Eds), *Ergonomics International '85*, pp. 520–2, London: Taylor & Francis Ltd.

Dray, S. M., 1987, Getting the baby into the bathwater: putting organizational planning into the systems design process, in Bullinger, H. G. and Schackel, B., (Eds), *Human Computer Interaction–Interact '87*, pp. 793–7, Amsterdam: Elsevier.

Dray, S. M. and Monod, E., 1986, The new 'internationalism' in macroergonomics, in Brown, O. and Hendrick, H. W., (Eds), *Human Factors in Organizational Design and Management II*, pp. 499–503, Amsterdam: Elsevier.

Emery, F. E. and Trist, E. L., 1960, Sociotechnical systems, in Churchman, C. W. and Verhulst, M., (Eds), *Management Science: Models and Techniques*, Vol. 2, pp. 83–97, Oxford: Pergamon Press.

Freedman, D. H., 1986, In search of productivity, *Infosystems*, **33**, 12–14.

Hendrick, H. W., 1986, Macroergonomics: a conceptual model for integrating human factors with organizational design, in Brown, O. and Hendrick, H. W., (Eds), *Human Factors in Organizational Design and Management II*, pp. 467–77, Amsterdam: Elsevier.

Hoerr, J., Pollock, M. A. and Whiteside, D. E., 1986, Management discovers the human side of automation, *Business Week*, **29 Sep.**, 70–5.

Kelleher, J., 1986, The transformed organization, *Information Center*, **26**, 31–5.

Kling, R. and Scacchi, W., 1980, Computing as social action: the social dynamics of computing in complex organizations, *Advances in Computers*, **19**, 249–327.

Lawler, III, E. E., 1986, *High-Involvement Management*, San Francisco, California: Jossey-Bass.

Lawler, III, E. E., Renwick, P. A. and Bullock, R. J., 1981, Employee influence on decisions: an analysis, *Journal of Occupational Behavior*, **2**, 115–23.

Leonard-Barton, D. and Kraus, W. A., 1985, Implementing new technology, *Harvard Business Review*, **85**, 102–10.

Likert, R., 1961, *New Patterns of Management*, New York: McGraw-Hill.

Mainiero, L. A. and DeMichiell, R. L., 1986, Minimizing employee resistance to technological change, *Personnel*, **46**, 32–7.

McGregor, D., 1960, *The Human Side of Enterprise*, New York: McGraw-Hill.

Mumford, E., 1983, *Designing Human Systems*, Manchester: Manchester Business School.

Mumford, E., 1987, 'XSEL — the participative design of an expert system, presentation at Interact '87', Straggart, Germany.

Pava, C., 1983, Managing new information technology: design or default? in Walton, R. E. and Lawrence, P. R., (Eds), *HRM Trends and Challenges*, pp. 69–102, Boston, Massachusetts: Harvard Business School Press.

Robertson, M. M. and Dray, S. M., 1988, Office automation in America: a systems perspective, in Adams, A. S., Hall, R. R., McPhee, B. J. and Oxenburgh, M. S., (Eds), *Proceedings of the Tenth Congress of the International Ergonomics Association*, pp. 732–4, Sydney: Ergonomics Society of Australia Inc.

Schelling, T. C., 1980, *The Strategy of Conflict*, 2nd Edn, Cambridge, Massachusetts: Harvard University Press.

Sullivan, D., 1986, 'The marriage of technology and hotel operations', keynote speech to Hospitality Tech. at the New York Hilton.

Taylor, J., 1984, *Participative Socio-Technical Design and Improving QWL, a Sampler of North American Socio-Technical Systems Cases*, pp. 78–105, internal document of Socio-Technical Design Consultants, Inc. (Available from Dr James Taylor, Human Factors Department, ISSM, University of Southern California, Los Angeles, CA 90089–0021, USA.)

Westin, A., Schweder, H., Baker, M. A. and Lehman, S., 1985, *The Changing Workplace: A Guide to Managing People, Organizational and Regulatory Aspects of Office Technology*, New York: Knowledge Industry, Inc.

Zemke, R., 1987, Sociotechnical systems: bringing people and technology together, *Training*, **24**, 47–57.

Chapter 10

Sailors on the Captain's Bridge: A Chronology of Participative Policy and Practice in Sweden 1945–1990

O. Östberg, L. Chapman and K. Miezio

Introduction

Sweden is a leading representative of progressive Scandinavian and European working life. Developments in Sweden between 1945 and 1990 illustrate the evolution of participatory structures in job ergonomics, task analyses, organizational methods, and national policy. Employee participation in Sweden has been embedded within a larger framework of social institutions for participative management and co-determination. Today, worker involvement is not limited to helping select more comfortable seating equipment and hand tools. Instead, employees at the firm level also take part in the development of work standards, production-process improvements, research agendas, and other activities traditionally reserved for management decision-making. The historical development and the successes and failures of these new types of work roles and organizational designs are presented here in three chronicles:

1. changes in economic and labour policy,
2. the evolution of an example manufacturing industry, and
3. participative innovations in service industries.

The Swedish labour movement has been unique among union organizations in other industrialized nations in its ability to shift the terms of debate toward democratization at various levels of the work organization and the national economy (Sirianni, 1987). Sweden is also unique for a number of reasons. Certain management incentives to innovate and adopt work reforms have existed in Sweden for decades.

Managers in the USA and many other industrialized economies have relied on high unemployment levels to provide replacement workers when faced with high turnover and employee dissatisfaction. In contrast, Swedish unemployment rates have ranged from 1.5 to 4.0 per cent in the post World War II period. These conditions of relative labour shortage have provided the impetus for Swedish managers to experiment with a variety of participatory reforms on a broad scale in order to attract and retain workers (Cole, 1987).

Sweden has also been forced to seek improvements in quality and productivity due to its long-standing reliance on export industries. In contrast, US businesses until recently have benefitted from much larger domestic markets and a lengthy period of predominance in international trade (Cole, 1987).

Work in Sweden is organized and employees have independent representation so they stand on a more equal footing with employers and government. The level of unionization has been in the region of 90 per cent in Sweden for many decades while in the USA unionization has declined from a high of 33 per cent in 1953 to near 14 per cent in 1985 (Farber, 1987).

Three lines of development in Sweden are reviewed in this chapter:

1. Changes in national policies that have influenced work roles and conditions since 1945;
2. developments in manufacturing work as represented by the automotive industry since 1960; and
3. the evolution of strategies for improving information work from 1973 to the present.

The lessons drawn from the history of Swedish work reform indicate that participation at work depends on empowering individual workers and independent employee organizations.

Progressing together: national level economic policy and framework legislation (1945–1982)

Five distinct periods can be identified in the democratization of working life in Sweden (see table 10.1). Each period can be characterized by the passage of framework legislation which sequentially increased the participatory rights and responsibilities legally assigned to the Swedish work-force.

Table 10.1. Changes in Swedish economic and labour policy.

Year	Change
1946	Conventional unionism
1952	Passive labour market policy
1966	Active industrial policy
1972	Work environment policy
1977	Codetermination of Work Act
1982	New factories concept

Conventional management–unionism (1946)

In the period following World War II, unions and management were keenly aware of the need for Sweden's small, open economy to remain internationally competitive (Cole, 1987). Both parties also shared an understanding of the crucial role technological innovation played in producing sustained and stable economic growth. As a result, Swedish unions largely subordinated the traditional labour ideology of advancing job autonomy and satisfaction in the interests of gaining the economic benefits available from rationalized mass production.

Trade-union policies during this period emphasized the negotiation of wage scales and basic contractual rights. On the national level, a gradual process of organizing and mobilizing the labour force, that had been well underway for over 30 years, continued. Levels of unionization in the period 1946–1960 reached and have since maintained levels of 90 per cent of the total Swedish work-force (somewhat lower for white-collar workers). Labour organizations became highly centralized into three confederations organized along industrial-sector lines, helping to consolidate union power as a political force. All of Sweden's private employers also formed a confederation. Employers in the federal government, local governments, and in cooperative firms also formed bargaining blocks.

Passive labour market policy (1951)

In the interests of promoting growth and competitiveness and fuller employment, Sweden embraced a national economic plan worked out in the late 1940s by union economists referred to as the Rehn-Meidner model (Landsorganizationen i Sverige (LO), 1953). The plan incorporated moderately restrictive government fiscal and monetary policies and a national system of standard union wages. The Rehn-Meidner model favoured low cost, competitive firms that were able to pay standard wage rates and expand out of profits. The model also provided incentives to other firms to improve efficiency or go out of business. The plan's intent was to use market forces at the enterprise level to equalize wages across and within occupations while fostering productivity growth and research and development. The policy was designed to operate without direct intervention by either the state or unions in the process of structural and economic change itself. Provisions were included to assist those in the work-force who lost their jobs as a result of the new policy by improving the quality and availability of information about new jobs and by increasing retraining and relocation assistance.

This was not a participative policy that involved organized labour and individual employees, but rather a hands-off, market-driven model. Following its implementation in 1951, a national consensus prevailed during the 1950s and early 1960s that the Rehn-Meidner model was producing enough jobs while at the same time helping to keep Sweden competitive by stimulating industrial efficiency and technological innovation. Support for the policy began to erode during the 1960s (Martin, 1987).

Active industrial policy (1966)

Although the Rehn-Meidner labour market policy model helped Sweden attain high levels of affluence and employment, by the mid-1960s greater competitive pressures were being felt from the world economy. Sweden's economy, with its limited domestic market and significant export dependence (40 per cent of industrial production) proved particularly sensitive to the less favourable external conditions that developed and to the accompanying pressures for structural changes. By 1968, economic growth in Sweden had become markedly more rapid and unstable. Communities and individuals were being adversely affected by the extreme geographic mobility of labour that was encouraged by the existing labour market policy. The Rehn-Meidner model involved wage-earner solidarity (salaries were not market-driven) coupled with provisions designed to support re-employment of displaced workers and to sustain local economies, but the model was overwhelmed by the speed with which job dislocation and job creation were taking place.

Two other factors contributed to the reshaping of the Swedish society and working life. One was the rapid influx of immigrant workers, workers who were to become part of the wage-earner solidarity. One result was a law requiring employers to provide Swedish language courses. More important was the ambition to increase the role of women in the work-force. This meant that Sweden had to design better financial and legal opportunities for a less gender-based family and working life, such as the creation of facilities for day care (sometimes also night care).

As a result, a more activist approach with an expanded role for the state in managing the economy was adopted with the intention of steering development 'so that it is of benefit to all'. Not only did the public sector grow. Planning documents stressed the importance of establishing the conditions necessary to make structural and technological change more compatible with security, justice, and environmental protection, particularly where the actions of private enterprise were liable to exact large, uncompensated social and human costs (LO, 1971). New governmental institutions that fostered greater planning, higher levels of macroeconomic coordination, and increased public investment in advanced technology were developed including a State Investment Bank, a Ministry of Industry, and a Board for Technical Development. However, labour's participation in the new industrial policy was minimal since decisions about economic and technological change remained managerial prerogatives beyond the scope of collective bargaining.

Work environment policy (1972)

In the late 1960s and 1970s, trade unions pressed a broad front of new framework legislation into being. The driving force was that the quality of working life had not kept pace with the increased standard of housing and living. Fundamental concepts embodied in these policies included:

1. more participatory structures that involved labour in both planning and remedial efforts in the form of decentralized workplace level institutions;
2. adopting a broader, better integrated vision of work's influence on health and well being; and
3. placing work life in a context which included the larger realm of everyday social and individual lives.

The historical factors helping to motivate the drive toward participation and improved working lives included the attention being given at that time to workplace democracy projects in Norway which had fostered the concept of the autonomous work group instead of hierarchical, managerial control of work (Thorsrud and Emery, 1968). Until 1972, issues concerning the organization of work and applications of technology were considered managerial prerogatives and the constitution of the Swedish Employers' Confederation (SAF) required that a clause to this effect be contained in all collective agreements. Swedish unions had also grown dissatisfied with the limited advisory role on the national level assigned to them regarding labour market and industrial policy.

Sweden's trade-union movement came under pressure from the rank and file as a result of an unprecedented series of well publicized wild-cat strikes in large, important, and highly visible companies (see e.g. Korpi, 1978). The industrial unrest began in the late 1960s and continued up until the mid-1970s and centred on disparities in power relationships at the level of the workplace and on employee dissatisfaction with existing labour laws and working environments.

The outcome was a rapid sequence of new labour policies beginning with the creation, in 1972, of the Swedish Work Environment Fund (Sandberg, 1982). The Fund was financed by an employer pay-roll tax which in 1988 amounted to 0.13 per cent of total wages and yielded a total annual budget of about $100 million (in 1988 US dollars). The Work Environment Fund supported scientific research, demonstration projects and trade union training activities. In the years since it was established, the Work Environment Fund has shifted its emphasis in demonstration projects from supporting a few high-visibility model programmes in favour of a larger number of lower-profile local efforts. Overall, the model programmes approach proved to be not as successful as had been hoped in stimulating new developments throughout Swedish industries.

In an attempt to put in place a more just balance of power between labour and management, Sweden's union confederations forced the enactment of national legislation in 1974 which placed worker representatives on corporate boards of directors. Four years later the Swedish Work Environment Act of 1978 became law. The Act provided legal authority for plant level participation and significantly expanded employee job control rights. In addition, the Work Environment Act extended authority for national policy into a number of new areas. New hazards to physical health accentuated by technological change such as sedentary work and repetitive motion problems were covered for the first time in the Act as were psychological stressors including job-related cognitive strain, monotony, work abstraction, and time and production pressures. Great emphasis was placed on problem prevention by forcing businesses to introduce work environment considerations early on, while equipment and organizational changes were still being planned. Finally, serious attention was given in policy statements to considering work environment implications in the formation of industrial policy, labour market policy, and structural change within industry. Particular emphasis was also placed on strengthening the position of local unions in pressing work environment demands. Labour's 1971 Work Environment Act policy statement concluded that 'the decisive question is to what extent the trade union movement can create the possibilities for influence by employees in the planning and decisionmaking' (LO, 1971).

Two new institutions were established at the plant level by the 1978 legislation: joint committees and the role of safety delegates. Firms with more than 50 employees were required to establish joint labour management committees for health and safety. Via the committees, the unions were given extended opportunities to influence and bargain with regard to plant facility expansions and new construction, industrial health and safety, and company health services.

Firms with five or more employees were required to hold employee elections for safety delegates, that is, the unions elected safety representatives. Safety delegates also had clearly defined rights to information and participation in planning as well as the authority to stop work in the plant if there was an imminent danger of serious injury.

In addition to providing legal authority and institutional vehicles at the plant level for employee participation in occupational health and safety, an intensive training programme for safety delegates was established by the Work Environment Act and paid for by monies from the Work Environment Fund. By 1982, over 350 000 workers had received over 40 h of training. In the initial years, training programmes emphasized legal aspects and nontechnical information about workplace hazards. More recently, emphasis has shifted to cover technical training as well and also includes the development of expertise in making and evaluating environmental measurements. The Working Environment Fund also provides all Swedish safety delegates with a free monthly journal.[*]

Codetermination of Work Act (1977)

From the end of World War II until the 1960s, virtually all methods that increased productivity were accepted in Sweden, as long as they did not immediately threaten the physical health of workers. Gradually it was realized that the traditional methods of productivity improvement had severe shortcomings. Increased wages and shorter hours available with traditional methods of work rationalization were failing to compensate for unwanted social and psychological detriments to the work-force.

Up to that point, the School of Scientific Management and the Human Relations Movement were the systems tools most widely used in Swedish and in most Scandinavian industry (Sandberg, 1982). These tools were often employed in parallel but with minimal coordination: e.g. human relations did not deal to any great extent with issues of technology and formal organization, just as scientific management generally neglected social and human empirical research aimed at resolving this conflict through the introduction of autonomous work groups and related forms of work organization. The Industrial Democracy Programme in Norway served as a starting point for integrating work improvement efforts (Thorsrud and Emery, 1968).

During the 1960s and 1970s, numerous company programmes were started in Swedish industry, and many enjoyed consultative support from the technical department of the Swedish Employers' Confederation, which embraced and pushed this new development. Others were supported by joint employer–union committees set up to promote experiments and to establish model projects. The field experiments had much in common with programmes later launched in other nations under the headings 'quality of working life', 'humanization of work' and 'quality circles'.

These programmes typically focused on shop-floor issues and were characterized by unusually high levels of cooperation between management and the union local. However, it gradually became clear that the employer side had never intended to let the union side be part of any codetermination efforts beyond the day-to-day details at the shop-floor level. The fact that the central trade unions found it difficult to establish a reasonable relationship with regard to these issues was one of the major factors behind the culmination of this era and new developments in the legislation regulating the Swedish industrial relations system.

New factories concept (1982)

The backbone of the work-life reforms of the 1970s was the Codetermination at Work Act, made effective in 1977. The act was extremely comprehensive industrial relations legislation

---

[*] For a free subscription of the annual English edition of this informative journal, write to: Working Environment International, Birger Jarlsgatan 122, 114 20 Stockholm, Sweden.

Table 10.2.   Excerpts from the 1982 Swedish Agreement on Efficiency and Participation.

Job enrichment
Goals and direction of development work. 'Developing and improving the efficiency of the firm, together with safeguarding employment, are matters of common interest to the company and its employees. The organization of the work, and the jobs of individual employees shall be designed so that the employees are given as engaging and stimulating a job situation as possible. To achieve this, the parties shall seek to reach a common standpoint about the ways in which regular development work shall be carried out.'

Influence over work
Developing the work organization. 'The company's strength and competence contributes to security and employment. In the constant development at all company levels, varied and developing working practices shall be aimed for, so that individual employees can become capable of accepting more demanding and responsible jobs. Developing the work organization can, for instance, include measures to improve productivity, the introduction of group work, group organization, job rotation or job enrichment; decentralization and delegation are of decisive significance. The employees shall be given opportunity to take part in designing their own job situations as well as the work of change and development that affects their own job.'

Active participation
Technical development. 'Day-to-day as well as more far-reaching technical modernization offers many opportunities that must be taken to enable the company to survive and achieve success. When technical change that involves major changes for the employees is being planned, the trade union organizations shall receive information on the consideration underlying the new technology and the technical, financial/economical, work environmental, and employment consequences that can be forseen and possibly make proposals for appointing project groups.'

Access to information
The company resources. 'Forecasts by company management shall be presented so that the trade union organizations at the company have opportunity to consider and evaluate jointly with management the company's market prospects, purchasing activities, competitive position, product development and production equipment, and the job security of the employees and their development in their jobs. Each individual employee possesses unique knowledge about his/her own job situation, and therefore has very considerable possibilities of proposing improvements to the benefit of the overall result.'

Joint activities
Codetermination forms. 'When participating in line negotiations, bipartite bodies and project work, the trade union representatives shall be given reasonable time to enable them to evaluate matters arising. Trade union members are entitled to participate during paid working hours for a maximum of five hours per year in trade union meetings at the workplace arranged by local trade union organizations on mattters that are concerned with relations with the employer or otherwise connected with the trade union activities at the company.'

Access to outside expertise
Employee consultants. 'The local trade union organizations are entitled to employ employee consultants for special tasks prior to coming changes that are of significance to the company's financial/economic position and the jobs of the employees. The purpose of this is to give the trade union organizations the opportunity to analyse the facts available and adopt a standpoint on the situation and the changes it will give rise to. Employee consultants may be outside experts or employees of the company. The compnay is responsible for the reasonable costs of a consultant, the scope, content and cost of whose work shall be specified. The local trade union organizations and the employer shall jointly sign an agreement with external employee consultants concerning secrecy.'

and opened up all areas of company life to trade-union influence. It was a framework law that was later followed by central and local agreements between employers and the trade unions. For private industry, the follow-up agreement was not concluded until 1982. Contributing to the delay was disagreement about the 'new factories' concept as conceived by the Swedish Employers' Confederation (Agurén and Edgren, 1980). From an international perspective, this concept was quite advanced and among other things stressed ergonomics solutions such as those found at the much touted assembly line at the Volvo plant in Kalmar (Gyllenhammar, 1977). However, the union side refused to participate since there was no room for union influence at the higher levels of decision-making in the envisaged new factories. The new-factories concept focused on participative ergonomics in the traditional sense, but did not recognize trade-union demands with regard to continuing participation that included problem-definition phases.

Eventually, in 1982, the parties signed a Development Agreement (SAF, 1984). This was an agreement on efficiency and participation issues, intended to create a framework for positive cooperation at the enterprise level between management, employees, and trade unions on matters that concerned both the progress of the company in general and on a day-to-day solution finding. Excerpts are listed in table 10.2.

Developments in the Swedish automobile manufacturing industry (1960–1990)

All blue-collar workers in the automobile industry belong to the Union of Metal Workers, a labour organization which has long supported the introduction of new technology. One clear example of union attitudes was provided in the mid-1970s during the early phases of robotization when a union spokesperson indicated that working environment arguments about health hazards alone would justify replacing every tenth industry job with a robot (Östberg, 1978). This positive attitude toward modernization has prevailed. In 1982 the president of the Swedish Union of Metal Workers rebuked groups asking for contract language containing the right to veto new technology. Instead he suggested that they seek a veto right on old technology.

Behind such confidence was the knowledge that as a result of the policy framework already in place, individual workers in Sweden have a great deal of employment security and, collectively, they have considerable influence on the design, implementation, and operation of new production systems. This influence is achieved by means of the network of framework laws and agreements in which the introduction of new technology is approached as part of a larger set of worker and union rights to constant improvement of the workplace.

Volvo's biotechnology (1961)

Ergonomics on a large scale was introduced in Sweden through the biotechnology concept implemented at the Volvo Penta plant for engine assembly in 1961 (Yllö and Sandén, 1971). A rigid assembly line production process, based on piece-work rates, paced-work, and methods-time measurement (MTM)-designed jobs, was modified in order better to consider human biological capabilities and limitations. The driving force for the biotechnology initiative was the plant's safety and health unit, which came to serve as a model for industrial ergonomics in Sweden. In an ambitious training programme, production engineers were taught biotechnology principles and were encouraged to bring in outside ergonomics expertise during major development projects. The basic goals were to aim for an 'optimum between risk and luxury' in working conditions and, at minimum, to make workplaces more physically safe and healthy.

Saab's sociotechnology (1970)

Volvo's biotechnology was very successful, but during the 1960s there was a realization that merely incorporating physical safety and health into work design was not enough. Problems remained, especially worker monotony and alienation, that continued to produce high turnover, absenteeism, and low productivity. The labour movement argued that it was not enough to have shorter working hours, higher wages, longer annual leave, better injury prevention, and more equitable worker compensation for illnesses and injuries. Better pay and physical health, fewer injuries, and more time off were still not enough to compensate for the detrimental psychological and physical effects of empty, meaningless and repetitive work. In fact, it became generally accepted that empty work resulted in empty leisure (Johansson, 1971).

The Saab-Scania assembly plant in Södertälje pioneered a sociotechnical approach as a means of enriching the working conditions. 'Socio' referred to work organization that made room for job enlargement, job rotation, social contacts, and some degree of self-determination within smaller production units. 'Technical' referred to refinements and new technology to boost productivity and develop production methods that uncoupled workers from the strict pacing of the assembly line.

At Saab-Scania, sociotechnology meant that parts of the production process were highly mechanized in traditional assembly-line fashion, while others emphasized human capabilities. In one example, a manual assembly line for building engines was broken up into parallel flow lines in the shape of loops on the main line (Goldman, 1979; Agurén and Edgren, 1980). Figure 10.1 depicts graphically what the changes in the new lines entailed. This arrangement was achieved by means of ergonomically designed assembly trolleys that could be individually pulled into the loops and assigned to one of the workers within the loop unit. With these trolleys, an engine block could easily be positioned at the correct working height for the workers, who could then carry out a complete assembly cycle. Most of the workers were women and, thanks to the loop system (which also served as an in and out buffer), they could help each other during difficult or heavy work tasks.

This system was highly productive and met some of the desire for quality of working life. But shortcomings existed with regard to both the 'socio' and the 'technical' subsystems. For example, one serious technical drawback with engine production was that the loops were still lines within which it was not possible to change position between individual assembly trolleys. Other problems arose. The work-force at the Söldertälje plant was made up of a diverse range of nationalities with a great many language barriers that proved to be obstacles to needed levels of social interaction.

Although no salary penalties were imposed if the production quota could not be met, the

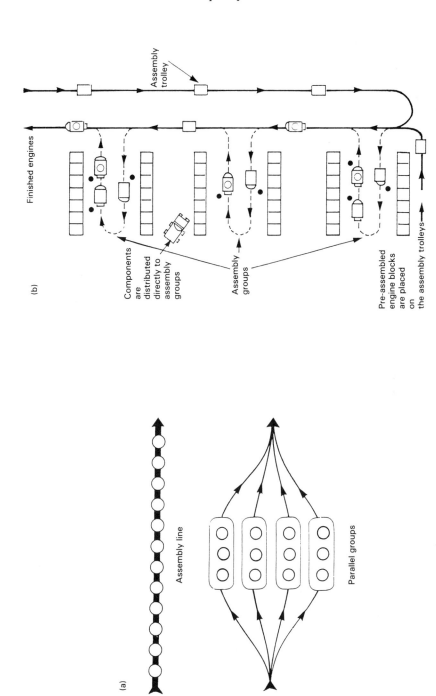

(b)

Finished engines

Assembly trolley

Components are distributed directly to assembly groups

Assembly groups

Pre-assembled engine blocks are placed on the assembly trolleys

(a)

Assembly line

Parallel groups

Figure 10.1. Assembly line versus parallel flow lines in Swedish automobile production. (a) An assembly line and parallel groups. (b) Assembly of gasoline engines at Saab-Scania in Södertälje; the work process in three of the parallel loops is shown. Reprinted from Agurén and Edgren (1980).

workers were asked to discuss and find solutions to the problems. This resulted in group pressure on workers who fell behind in production; it also caused the workers in a loop to speed up production voluntarily in order to obtain a period of non-work of up to 2 h at the end of the shift. Overall, more pressure was exerted for greater production by workers on each other than had formerly been required by management.

The work-load under the group-paced assembly production was witnessed by a group of American automobile workers evaluating the Saab-Scania system. As reported by Goldman (1979), the American workers found the plant superior to American plants with regard to ergonomics and environmental factors, but they were troubled because the women 'worked harder than anyone in Detroit'. This hard work showed up in more sick leave taken by the Saab-Scania workers. In addition, the company physiotherapists were busy treating neck, shoulder, and arm ailments among the assembly workers.

Volvo's teamwork (1974)

The Volvo Skövde plant for engine assembly, which opened in 1974, amplified both the biotechnology and sociotechnology concepts. The production system had vertical job enlargement, self-propelled assembly carts, ergonomically designed work stations and hand tools, and a flow layout and plant architecture that supported the team-work concept. The Volvo Kalmar plant for final car assembly has become better known internationally and is based on the same principles. Union and management representatives from all over the world have made study visits to the Kalmar plant during the last decade.

Management described the Kalmar plant as the most productive within the Volvo family. Though the workers were positive about the basic ideas, they maintained that the plant has received more praise than it deserves. Their main criticism was that production is still based on a moving-belt technology, even though the robot carts make the belt invisible. They also felt that there was not enough buffer capacity in the loops, and that it takes young, strong workers to manage the physical demands posed by the production speed. In the years since the plant's opening, production streamlining and higher utilization of the production resources have eroded some of the initial worker freedom (Agurén *et al.*, 1984). Still, in 1988, a television commercial in the USA used scenes from the Volvo Kalmar plant to promote the concept of happy workers making good cars.

Saab's self-paced work (1983)

While the world was still admiring the teamwork in Swedish automobile assembly plants, the Saab Scania plant for engine assembly in Södertälje went further and actually abandoned the assembly line, moving belt, and teamwork concepts (Östberg and Enqvist, 1984). The work design principles of biotechnology and ergonomics were further amplified here, as was the drive toward truly parallel production flow lines. A crucial advance was the development of a computer-aided assembly carrier that surpassed Volvo's robot cart.

In the new Saab Scania system, a number of heavy or short cycle operations were assigned to industrial robots. Their tasks included mounting valve springs, sealing valves, mounting, flywheels, and threading and torquing bolts. Great care was taken to separate the robot and the manual assembly operations. The manual assembly was carried out in individual, ergonomically well-designed work stations. As the workers completed their tasks, they sent away finished products and ordered new assembly carriers according to their individual workpace.

Purely ergonomic considerations aside, the new assembly system was launched primarily to boost output capacity. In particular, it was designed to handle custom-order production. Thus about 30 different engine versions were being assembled at any given time. For the workers, this self-paced work was an improvement over the previous group-paced work, with its concomitant, self-inflicted high work-load. Even in cases where overall production was greater with the new system, self-paced work was less stressful than that set by groups of workers.

Saab's human centred production matrix (1987)

The prevailing impression from Saab's self-paced, robot-equipped Scania plant has been described as 'Good-bye to group assembly. No more stress and load injuries — but a less interesting job' (Enqvist, 1984). To put the interest back into the jobs, the Saab Trollhättan plant for car bodies resurrected the work-team idea and revitalized the concepts of job rotation and job enlargement. Although the concepts were first implemented at the Volvo Kalmar plant, later adaptations of the production process at Kalmar eventually took precedence over the team-work and worker freedom. The new Saab Trollhättan plant seems to have solved some of the problems by organizing production in the form of flow-group production matrices with more independent production units. Figure 10.2 contrasts traditional methods of organizing by functional grouping versus the greater independence and enlarged tasks possible with flow grouping. Flow groups can also reduce administrative requirements so that a single order form is sufficient for an entire manufacturing cycle. Of utmost importance at Trollhättan were the 1-h work buffers between each production unit.

The Saab Trollhättan plant's 7200 production workers are typically found in teams consisting of two women and eight men responsible for the team's output through:

1. assembly and materials handling;
2. service and maintenance;
3. quality control and adjustments; and
4. administration and planning.

There are three salary levels; after having received the necessary training, and contingent on management approval, the team decides when a member is ready to take on new responsibilities.

To support vertical job enlargement into the planning and quality-control functions, personal computers (Macintosh 512 Plus model, Apple Computer, Inc., Cupertino, CA) are available to the production teams. This support was initiated by a 5-year development programme, set up in 1982 to further the spirit of the Agreement on Efficiency and Participation signed that year between the main social partners in the Swedish labour market. The programme was financed by the Working Environment Fund (Arbetsmiljöfonden (AMF)), which has also published a report on the Saab Trollhättan experiences (AMF, 1988).

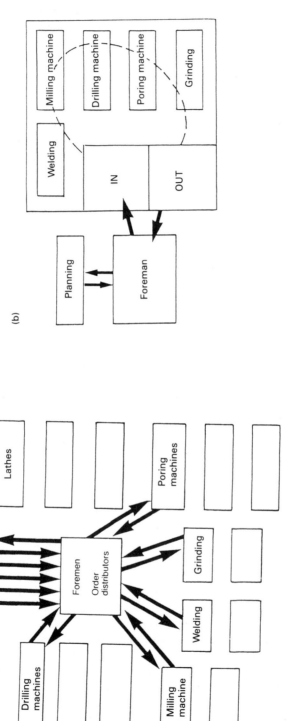

Figure 10.2.   Functional grouping versus flow group matrices in Swedish automobile production. (a) Planning process in functional machine grouping; it is necessary to allocate orders and to make out separate order forms for each individual operation. (b) Planning can be simplified through the use of flow groups. Here the entire group is a planning point and a single order form is sufficient for an entire manufacturing cycle. Administration is reduced and, in principle, no operational planning is needed beyond planning supply of materials to the group. Day-to-day planning and distribution of the work can be done within the group. This means that those who work in a flow group can obtain enlarged tasks and greater independence in their jobs. Reprinted from Agurén and Edgren (1980).

Statistically speaking, the Saab production work now consists of 40 per cent traditional manual line assembly, 30 per cent manual line service (housekeeping, preventive maintenance, magazine loading, etc.), 15 per cent manual work in parallel flow lines, 10 per cent work in large matrix units in complex transfer lines, and 5 per cent in small robot-based, flexible manufacturing system (FMS) matrix units. No statistics are available for individual workers, but there is a true job rotation within the work teams and a true differentiation in skill and responsibility demands between the various job functions.

For once the unions seem to be satisfied with the organization of production. The blue-collar workers' local union holds that 95 per cent of all production decisions are made in the production teams and in close cooperation with the management. The Central Organization of Salaried Employees in Sweden (Tjänstemännens Centralorganization (TCO)) was so impressed by the results of the Saab Trollhättan development that in 1987 the plant manager was given the prestigious TCO Award for Computerization. The manager has since been promoted to help implement the Trollhättan principles throughout the Saab Corporation. The previous year the award was given to a safety representative fighting for safety and health issues concerning visual display terminal (VDT) work.

A sad epilogue may be forthcoming. The Saab Automobile Division has lately experienced serious economic problems. As a result, as from 1989, the majority owner is General Motors. The new owner has signalled that traditional moving belt production may have to be reintroduced as a means of cutting down the production costs. Similarly, Volvo, also undergoing a bit of recession, in 1990 decided that the experimental 'whole truck assembly' would not be made a full-scale operation.

Service sector and information work developments in Sweden (1973–1988)

Regulatory policy, product standards and procurement guidelines: computer work station equipment (1973–1987)

By the mid-1980s, Sweden led the world in the number of industrial robots and bank office VDTs per capita. Overall, about one in four of Sweden's work-force were daily computer users (SCB, 1985). Swedish researchers were among the first to recognize the crucial health implications of computer equipment and computer working environments and task designs.

In 1973, Sweden enacted regulations to protect against threats to individual privacy posed by the increasing use of electronic data bases. At the same time, early investigations of VDT operators sponsored by Swedish unions and the Work Environment Fund identified high rates of eye strain and musculoskeletal ailments attributable to deficiencies in VDT equipment design, and in the lighting of the work environment, in the work content, and in the length of work shifts (Östberg, 1975; Gunnarsson and Östberg, 1977). Given Sweden's policy framework, it was then natural for unions and governmental officials to turn their attention to technology-forcing standards to improve the physical ergonomics of the human–computer interface. The initial strategy utilized governmental regulations and a set of binding rules regarding VDT screen characteristics affecting Swedish employers was passed in 1978.

A more comprehensive, yet also more flexible, ordinance has since replaced the 1978 rules which covers work design features as well as equipment ergonomics (Arbetarskydsstyrelsen (NBOSH) 1985). The new regulation also uses performance-based criteria and so is more flexible than rules depending on exhaustive specifications. In addition, the Swedish government has developed national standards for electromagnetic-radiation-measurement procedures and operates a comprehensive testing programme through an underwriter authorization system (Statens Mät-och Provstyrelse (MPR) 1987).

Although Sweden's rates of technological innovation and computer use are among the highest in the world, most computer equipment is imported and computer-equipment manufacturers have been slow to respond to Swedish ergonomic standards. To stimulate the evolution of technology, the development and widespread distribution of procurement guidelines is being given greater attention and may ultimately prove to be more effective than legislation (Styrelsen för teknisk utveckling (NBTD) 1986). Sweden's largest employer and VDT user, the Swedish Telecommunications Administration, first developed a national procurement policy for VDTs in 1982 (Östberg *et al.*, 1986).

Participatory research: the Nordic printing industry's UTOPIA Project (1980–1986)

To participate fully in the development of new technologies and work-life improvements, workers need access to both information and expertise. The Swedish Trade Union movement has long recognized that where experts stand on scientific and technical issues depends on where they sit. Until the early 1970s, research and development advice about work improvements nearly always came from professionals working for management (Sandberg, 1983). In 1978, the Swedish Working Environment Act incorporated innovative provisions that provided funding for unions to hire their own scientific help and participate from the beginning in joint research projects.

In 1981, Sweden's blue-collar unions produced a comprehensive report on research policy (LO, 1981). The report noted that the ability of a trade union to influence technological change and work-life improvements remained limited to the degree that labour could:

1. determine the questions and define the problems that were the object of research, and
2. bring scientific resources to bear on them.

Simply extending participation to include joint labour—management research efforts were seen as inadequate since it precluded research that employers did not regard as being in their interest. The largest obstacle to autonomous union research was perceived to be lack of resources. This led to demands for greater public funding and the development of union capacities to direct and administer scientific efforts as part of national research policy (Martin, 1987).

One result was a project to develop newspaper production technology. In the Scandinavian languages, UTOPIA is an acronym for 'training, technology, and products from a skilled worker's perspective'. The skilled workers in question were the members of the Nordic trade unions in the printing industry, and the project grew out of action-oriented research on behalf of organized labour (Arbetslivscentrum (ALC) 1981; Ehn and Sandberg, 1983). The UTOPIA project provided a clear example of the importance of the re-emerging craft model of work instead of specialization in job design (Sirianni, 1987).

Table 10.3. Aims of the Nordic printing industry's UTOPIA project.

1. Contribute toward the establishment of a Nordic trade union institute for the graphics industry
2. Delineate the technological/political strategies for the trade union movement
3. Develop work procedures for trade union involvement in technological development
4. Develop trade union alternative systems for text and image processing
5. Develop trade union training programmes for text and image processing

When UTOPIA became a public funded, independent development project in 1981, its 3-year research effort had five objectives (listed in table 10.3). A central view of the project was that, with existing technology, there were limited possibilities for local influence on work organizations and on the quality of work and products. The logical approach was then not to accept the technology as given. At the same time, trade unions recognized that alternative technologies must be technical successes and allow firms to remain competitive in international markets. To achieve this goal the Nordic graphics workers sought the help of experts. Union resources together with the public funding made available expert advice in areas such as computer sciences, sociology, and ergonomics.

The evolving 'union label' technology attracted interest from industry. After about 18 months the UTOPIA project was approached by Liber, a Swedish state-owned printing concern, and a cooperative agreement was signed to the effect that UTOPIA should furnish 'applications specifications' that would later be incorporated into Liber's new generation of printing equipment, planned to be a total investment of US$10 million. The applications specifications dealt with ways for graphics workers to continue to play a significant role in the printing process. A typical technical/economic problem was finding a balance between workers' demands to be able to see a full newspaper screen displayed at once despite screen-technology limitations. Eventually, the UTOPIA requirements covered the areas listed in table 10.4.

The UTOPIA development work turned out to be a technical success in that the Liber system was a success. Under the trademark TIPS, Liber had sold nine systems by the end of 1984. UTOPIA is keenly following the gradually phasing in of the new TIPS systems. Customer demands have resulted in certain compromises with the original UTOPIA specifications. At one Swedish newspaper, however, the local journalists' union, wary of

Table 10.4. Areas covered by the requirements of the Nordic printing industry's UTOPIA project.

Page make-up	*Work organization*
Work procedures	Work-station ergonomics
The display screen	Work-equipment ergonomics
Interaction devices	Visual strain abatement
Operations	
Image processing	*Human–machine interaction*
Scanner	Advanced cognitive ergonomics
	Training
	Graphic worker 'computer sciences'

what its members saw as 'graphics workers' technology', refused to grant the UTOPIA team any role during the implementation of the new TIPS system (Howard, 1985). UTOPIA was a forerunner of what is today called 'desk top publishing'. The project team has summarized its experiences in a book entitled *Design of Work* (Greenbaum and Kyng, 1990).

Nation scale participatory ergonomics: the screen checker (1986)

Swedish graphics workers belong to the LO family of blue-collar unions while journalists belong to the Central Organization of Salaried Employees (TCO). In 1986, the TCO released the Screen Checker, a six-page check-list for quickly assessing the ergonomic features of computer keyboards and display screens. It was made available in the Scandinavian languages as well as in other major world languages including Japanese (TCO, 1987). The Screen Checker did not represent a major new development but instead brought together previous work in a form easily applied by end-users.

The Screen Checker has proven to be an influential tool for many VDT users. Its power was a combination of its timing, simplicity, and organizational back-up. For many years TCO had been pushing for research and development of safe, user-friendly, and ergonomically well-designed VDTs (Bovie and Östberg, 1983). After a series of less successful initiatives through the traditional political channels, TCO decided to assume the role of a primary force by mobilizing and empowering its members. The Screen Checker was introduced during the International Conference on Work With Display Units held in Stockholm, Sweden, in May 1986. Within 12 months some 100 000 copies had been distributed within Sweden. The check-list has already had a significant effect on organizations using VDTs, and on manufacturers of VDTs, and the governmental bodies responsible for the working-environment aspects of VDT work.

Responding in part to employee assessments utilizing the Screen Checker, the Swedish Association for Suppliers of Computers for Offices (LKD) issued an ergonomics fact sheet addressing many of the same questions that were addressed in the TCO check-list. The Swedish National Council for Metrology and Testing (MPR, 1987) established new authorization rules for ergonomics testing of VDTs in underwriters laboratories. The Swedish government has also postponed decisions regarding mandatory testing of VDT equipment in the hope that industry would voluntarily evaluate their products and publish the information. Even a voluntary testing recommendation might be seen as a 'trade barrier' within the European Community, if vigorously demanded by the trade unions.

The Screen Checker conformed to the general Swedish consensus on VDT design and usage, and did not introduce any new ergonomics or safety and health issues. The most innovative aspect of the Screen Checker was that hundreds of thousands of TCO-organized employees suddenly had access to a tool that allowed them to evaluate their work equipment according to recognized standards. They passed their findings on to their employers, their union, safety professionals, and equipment vendors and took part in a nationwide participatory ergonomics movement. In 1991, the successful Screen Checker was followed by the even more ambitious Software Program Evaluation Kit.

The future: TCO union policy statement (1988)

Sweden's Central Organization of Salaried Employees (TCO) represents most of the nation's service industry and especially information work employees. The TCO is committed to promoting the evolution and improvement of work life in service industries. A 1988 TCO report entitled 'Codetermination and the Politics of Work' reflected on trade union experience from previous years and charted the union's future course (TCO, 1988). The report interpreted Sweden's experience as illustrating that it is not possible to obtain democracy from within a firm. For example, it is not possible to entrust the democratization process to a Japanese-style union local, that is, a union local living in symbiosis with the employer.

There are many uncertainties in organizations, and they cannot be managed democratically if only inbred solution techniques are employed. The argument resembles the popular view that it is not possible for the fish in the water to imagine a life outside the water or the mathematical principle that it is not possible to prove all of an axiomatic system's theorems from within the very same system (Gödel, 1962). The law of requisite variety in cybernetics theory similarly states that it is not possible to steer a system having some inherent variety unless the steerer has access to even more variety (Ashby, 1958). That is, variety can only be handled successfully if there is an equal amount of variety in the approach that is used to manage it.

The Swedish (Scandinavian) approach to this codetermination dilemma has been to structure the situation through a legislative framework in combination with collective bargaining. At the same time it has been a holistic and three-pronged approach aiming at

1. workplace democracy,
2. physical safety and health, and
3. the quality of psycho-social work life.

The TCO report also described five changes that may threaten the continued influence of Swedish work reforms. The first change noted that the current legislative framework in Sweden was based on concepts relevant to a 'smoke stack' type of industry and, in particular, on the existence of a techno-bureaucracy with a clear-cut decision hierarchy. However, due to thinner layers of middle management and more use of computers, modern organizations have smaller decision pyramids. A second change involved franchising, compartmentalization and the creation of companies within companies. This included situations where production is seen as the coordination of different companies for owning the physical plant, hiring the work-force, owning the production machinery, and handling customer relations. A third factor of importance was the growing focus on control via a corporate culture based on strong and charismatic leaders. A fourth factor was market-driven production (be it in industry or not), whereby decisions are made on the spot and on demand. A fifth factor was the transnationalization of production, which is a significant factor in Sweden where 40 per cent of industry production is exported.

The TCO report concluded that the present system of industrial relations in Sweden is based on an outdated view of the *structure* of the decision-making, and that society and trade unions now have to pay closer attention to the *process* of the decision-making. This means that employee participation in *design* processes will increasingly have to involve participants taking an active part in the *decision* process, while at the same time having an external knowledge back-up similar to that of the employer. Eventually this can lead to increased trade union involvement in research and development, because that is where the decision process starts. This reorientation is escalating and the Scandinavian working life movement is still very much alive.

The transnational relevance of Swedish developments

Given that the Nordic approach to participation has led to end results admired by trade unions internationally, are sweeping reforms like those that have happened in the Nordic countries possible elsewhere? More specifically, could what has happened in Sweden become a reality in the USA? At least three major obstacles are present that currently prevent Sweden's experience from being directly transferable to the USA.

Cultural values

The societal context for change is qualitatively and quantitatively different in Sweden. As in the USA, the great majority of businesses in Sweden, nearly 90 per cent, are privately owned. Unlike the USA however, Sweden is a small homogeneous nation, secular and pragmatic, where the belief in state-sponsored social reform runs deep (Lohr, 1987). Sweden's experience has shown that increased social regulation of the economy, the workplace, and public health produce real benefits in the lives of individuals and in the society and national economy.

Cultural values in the USA place great emphasis on individualism and competitive struggle, and adversarial and hierarchical workplace relations have been one result. In the USA there is a fear of regulation of the workplace — a fear that, on balance, social and individual freedoms would be compromised rather than improved by social regulation.

Although Americans see themselves as charitable and willing to devote time and material wealth for the good of the community, many find it difficult to cooperate due to perceived fears that their ability to get ahead and move up the socio-economic ladder may be compromised (Bellah, 1985). As a consequence, useful strategies for individual empowerment and social benefit in the USA remain unexplored.

Sweden values stability in economic development and job security. However, social and economic institutions in the USA encourage risk taking for short-term gains. Some examples of this predisposition are American labour force policies, the national tax structure, and business investment and other economic development practices.

Levels of employment security and labour representation

Less job security and lower levels of unionization in the USA coincide with a situation where there is very little in the way of framework legislation resembling the policies in place in Sweden. Analysts in the USA have pointed out that lower levels of respect for the value of human resources have contributed to their mismanagement. A 1988 report by the Massachusetts Institute of Technology Commission on Industrial Productivity noted that the USA clings to certain practices that may have served the country well in the past when the American economy dominated world trade, but which are not serving the nation well now.

They cite maladapted human-resource policies as a principal cause of the decline in American industrial productivity. The report's authors claimed that the USA is the easiest country in the developed world in which to hire and fire people, e.g. people can be fired at will. In France, for example, workers receive one year's notice before they are dismissed (Anon., 1988).

Overtly antiunion actions by employers have become socially and politically acceptable in the USA including active resistance to organizing efforts and plant siting in locations unsympathetic to unions (Farber, 1987). The growth of managerial power in the USA increasingly threatens employee rights to make a free decision regarding collective organization. The numbers of unfair labour practice claims per election nearly doubled between 1974 and 1984 and the number of claims found to have actual merit in the USA National Labour Relations Board increased even more rapidly (Farber, 1988). In October 1984, a report by the US Congressional Subcommittee on Labour Management Relations concluded that 'Workers who favor unionization are not protected from retaliation and discrimination, but instead are subject to intimidation and unequal treatment, frequently losing their jobs for no other reason than their preference to be represented by a union' (Kelber, 1987).

Management–labour balance of power

The work-force in Sweden shares power with employers by virtue of an extensive set of framework laws and the substantive political influence of labour organizations. In the USA, an unequal distribution of power within the workplace has been called the fundamental obstacle to establishing national models and policies for meaningful worker participation programmes that contribute to greater productivity (Levitan and Johnson, 1982). In the USA, management thinking sees the retention of control and final decision-making authority as essential to profit maximization. American managers are now seeking additional productivity improvements through further extensions of managerial power into areas where labour has traditionally exercised some control such as job assignments, production standards, and crew sizes.

The wide salary disparity between managers and employees in the USA also reflects the unequal distribution of power. The USA manager–production worker differential far exceeds the pay differences in comparable situations in Sweden, West Germany* or Japan. Managerial power is also reflected in the layers of organizational bureaucracy that effectively segregate American workers from managers to a greater degree than in Sweden, West Germany or Japan (Levitan and Weneke, 1984). In Sweden, employer investments in training are substantial and include all segments of the work-force. In the USA, training investments are proportionately smaller and most employee training resources are expended on managerial and technical personnel.

In Sweden, cooperation works because employees and their representatives have real power. Worker participation in Sweden is characterized by employee control over programme budgets for training and consultation, far-reaching rights regarding planning for

---

* Publisher's note: this paper was written before the unification of Germany and relates to the area that was formerly West Germany. In this respect, the name West Germany has been retained.

changes in production methods and new processes and construction, and by the ability to refuse unsafe work and shut down dangerous operations. Swedish unions have access to employer financial data and can influence directly the focus of governmental research programmes on work and the work environment.

The forms that worker participation efforts have taken in the USA almost never include independent power to order industrial hygiene monitoring, hire consultants, or to make changes in work practices, programme budgets or other areas where operational influence is wielded by Swedish workers (Boden *et al.*, 1984). Even those American corporations which have established worker participation or related programmes use them sparingly and involve only a fraction of their employees (NYSE, 1982).

Despite these three substantial current obstacles, Sweden's example remains relevant and unsettling for the USA. Sweden's experience amounts to a case history proving that large-scale unionization can greatly advance worker participation and lead to real benefits for society as a whole. Advancing worker rights through a legislative framework in combination with collective bargaining and the development of labour-controlled workplace and social institutions has not produced economic ruin, a bureaucratic nightmare, or a confining leash on individual initiative. In Sweden the result has been unprecedented levels of democracy, an enviable standard of living, greater health and a higher quality of life. The most fundamental question about Sweden's relevance may involve asking 'will the USA long remain satisfied with less?'.

References

Agurén, S. and Edgren, J., 1980, *New Factories*, Stockholm: Swedish Employers' Confederation.

Agurén, S., Bredbacka, C., Hansson, R., Ihregren, K. and Karlsson, K. G., 1984, *Volvo Kalmar Revisited: Ten Years of Experience*, Stockholm: *SAF-LO-PTK*, Efficiency and Participation Development Council.

ALC, 1989, *The UTOPIA Project*, Stockholm: Centre for Working Life.

AMF (Arbetsmiljöfonden), 1988, *Man and Technology in Cooperation: Body Assembly at Saab Trollhättan*, (in Swedish), Stockholm: Working Environment Fund.

Anon., 1988, Report pinpoints ills of U.S. companies, *The Capitol Times*, **16 Feb.**

Ashby, W. R., 1958, *An Introduction to Cybernetics*, London: Chapman and Hall.

Bellah, R., 1985, *Habits of the Heart: Individualism and Commitment in American Life*, Berkeley: University of California Press.

Boden, L. I., Hall, J. A., Levenstein, C. and Punnett, L., 1984, The impact of health and safety committes, *Journal of Occupational Medicine*, **26**, 829–34.

Boivie, P. E. and Östberg, O., 1983, Programme of data policy for the Central Organization of Salaried Employees in Sweden (TCO), in Briefs, U., Ciborra, C. and Schneider, L. (Eds) *Systems Design For, With, and By the User*, p. 289–92, Amsterdam: North Holland.

Cole, R. C., 1987, The macropolitics of organizational change: a comparative analysis of the spread of small group activities, in Sirianni, C. (Ed.) *Worker Participation and the Politics of Reform*, pp. 34–66, Philadelphia: Temple University Press.

Ehn, P. and Sandberg, Å., 1983, Local union influence on technology and work organization, in Briefs, U., Ciborra, C. and Schneider, L. (Eds) *Systems Design For, With, and By the User*, pp. 427–37, Amsterdam: North Holland.

Enqvist, J., 1984, Good-bye to group assembly, *Working Environment in Sweden*, **7**, 14–17.

Farber, H. S., 1987, The recent decline of unionization in the United States, *Science*, **238**, 915–20.

Farber, H. S., 1988, The union movement (letter), *Science*, **239**, 127–8.

Gödel, K., 1962, *On Formally Undecided Propositions of Principia Mathematica and Related Systems*, (translation of Gödel's 1930 treatise), London: Oliver and Boyd.

Greenbaum, J. and Kyng, M. (Eds), 1990, *Design at Work*, Baltimore: Lawrence Erlbaum Associates.

Goldmann, R., 1979, Six automobile workers in Sweden, in Schrank, R. (Ed.) *American Workers Abroad*, pp. 15–55, Cambridge, Massachuesetts: MIT Press.

Gunnarsson, E. and Östberg, O., 1977, *The Physical and Psychological Working Environment in a Terminal-Based Computer System*, (in Swedish), Stockholm: National Board of Occupational Safety and Health.

Gyllenhammar, P. G., 1977, How Volvo adapts to people, *Harvard Business Review*, **55**, 102–13.

Howard, R., 1985, UTOPIA: where workers craft new technology, *Technology Review*, **April**, 43–9.

Johansson, S., 1971, *Investigation of the Standard of Living*, (in Swedish), Stockholm: Allmänna Förlaget.

Kelber, H., 1987, The pressures against joining unions (letter), *New York Times* 25 Dec.

Korpi, W., 1978, *The Working Class in Welfare Capitalism*, London: Routledge and Kegan Paul.

Levitan, S. A. and Johnson, C. M., 1982, *Second Thoughts on Work*, pp. 173–99, Kalamazoo: Upjohn Institute for Employment Research.

Levitan, S. A. and Weneke, D., 1984, Worker participation and productivity change, *U.S. Department of Labour Monthly Labor Review*, **Sep.**, 28–33.

LO, 1953, *The Trade Union Movement and Full Employment*. Stockholm: LO.

LO, 1971, *Industrial Democracy*, Stockholm: LO.

LO, 1981, *Trade Union Movement and Research*, (in Swedish), Stockholm: LO.

Lohr, S., 1987, Swedes instill a sense of responsibility, *New York Times Fall Review of Education*, **30 Aug.**, 19.

Martin, A., 1987, Unions, the quality of work and technological change in Sweden, in Sirianni, C. (Ed.) *Worker Participation and the Politics of Reform*, pp. 95–139, Philadelphia: Temple University Press.

MPR, 1987, *Testing Visual Display Units — Test Methods*, (*MPR-P 1987:2*), Borås: Swedish National Council for Metrology and Testing.

NBOSH, 1985, *Ordinance Concerning Work with Computer Displays*, Stockholm: National Board of Occupational Safety and Health.

NBTD, 1986, *Technology Purchasing for the Disabled in Working Life*, (in Swedish), Stockholm: National Board for Technological Development.

NYSE, 1982, *People and Productivity: A Challenge to Corporate America*, New York: New York Stock Exchange Office of Economic Research.

Östberg, O., 1975, CRTs pose health problems for operators, *International Journal of Occupational Safety and Health*, **Nov/Dec.**, 24–6, 46, 50, 52.

Östberg, O., 1978, *Production with Limited Staffing: How? What will the Social Consequences be?* (in Swedish), Luleå: University of Luleå (1978:50T).

Östberg, O. and Enqvist, J., 1984, Robotics in the workplace: robot factors, human factors, and humane factors, in Hendrick, H. W. and Brown, O. (Eds) *Human Factors in Organizational Design and Management*, pp. 447–60, Amsterdam: Elsevier.

Östberg, O., Möller, L. and Ahlström, G., 1986, Ergonomic procurement guidelines for visual display units as a tool for progressive change, *Behavior and Information Technology*, **5**, 71–80.

SAF (Svenska Arbetsgivarföreningen), 1984, *Progressing Together: Developmemt Agreement*, Stockholm: Swedish Employer's Confederation.

Sandberg, Å., 1982, *From Satisfaction to Democratization*, Stockholm: Centre for Working Life.

Sandberg, Å., 1983, Trade union oriented research for democratization of planning in work life: problems and potentials, *Journal of Occupational Behaviour*, **4**, 59–71.

SCB (Statistiska Centralbyrån), 1985, *The Computer in Sweden: Survey on Computer Use, June 1984*, (in Swedish), Stockholm: Central Bureau of Statistics.

Sirianni, C., 1987, Worker participation in the late twentieth century: some critical issues, in Sirianni, C., (Ed.) *Worker Participation and the Politics of Reform*, pp. 3–33, Philadelphia: Temple University Press.

TCO, 1987, *Screen Checker: for Checking Computer Terminals, Word Processors and Personal Computers*, Stockholm: TCO.

TCO, 1988, *Codetermination and the Politics of Work: Strategies and Ideologies in Modern Working Life*, (in Swedish), Stockholm: TCO.

Thorsrud, E. and Emery, F. E., 1968, *Form and Content in Industrial Democracy*, London: Tavistock.

Yllö, A. and Sandén, S., 1971, *Biotechnology Checklist*, 3rd Edn, (in Swedish) Stockholm: PA Rådet.

Index